AF388980

Modelling of River Flows, Sediment and Contaminants Transport

Modelling of River Flows, Sediment and Contaminants Transport

Editor

Bommanna Krishnappan

MDPI • Basel • Beijing • Wuhan • Barcelona • Belgrade • Manchester • Tokyo • Cluj • Tianjin

Editor
Bommanna Krishnappan
Design
Krishnappan Environmental
Consultancy
Hamilton, Ontario
Canada

Editorial Office
MDPI
St. Alban-Anlage 66
4052 Basel, Switzerland

This is a reprint of articles from the Special Issue published online in the open access journal *Water* (ISSN 2073-4441) (available at: www.mdpi.com/journal/water/special_issues/river_flows_sediment).

For citation purposes, cite each article independently as indicated on the article page online and as indicated below:

LastName, A.A.; LastName, B.B.; LastName, C.C. Article Title. *Journal Name* **Year**, *Volume Number*, Page Range.

ISBN 978-3-0365-3530-2 (Hbk)
ISBN 978-3-0365-3529-6 (PDF)

Contents

About the Editor . . . . **vii**

Preface to "Modelling of River Flows, Sediment and Contaminants Transport" . . . . **ix**

Bommanna G. Krishnappan
Modelling of River Flows, Sediment and Contaminants Transport
Reprinted from: *Water* **2022**, *14*, 649, doi:10.3390/w14040649 . . . . **1**

Micheal Stone, Bommanna G. Krishnappan, Uldis Silins, Monica B. Emelko, Chris H. S. Williams and Adrian L. Collins et al.
A New Framework for Modelling Fine Sediment Transport in Rivers Includes Flocculation to Inform Reservoir Management in Wildfire Impacted Watersheds
Reprinted from: *Water* **2021**, *13*, 2319, doi:10.3390/w13172319 . . . . **5**

Dongdong Jia, Jianyin Zhou, Xuejun Shao and Xingnong Zhang
3D Numerical Simulation of Gravity-Driven Motion of Fine-Grained Sediment Deposits in Large Reservoirs
Reprinted from: *Water* **2021**, *13*, 1868, doi:10.3390/w13131868 . . . . **33**

Pankyes Datok, Sabine Sauvage, Clément Fabre, Alain Laraque, Sylvain Ouillon and Guy Moukandi N'kaya et al.
Sediment Balance Estimation of the 'Cuvette Centrale' of the Congo River Basin Using the SWAT Hydrological Model
Reprinted from: *Water* **2021**, *13*, 1388, doi:10.3390/w13101388 . . . . **49**

Bommanna G. Krishnappan
Review of a Semi-Empirical Modelling Approach for Cohesive Sediment Transport in River Systems
Reprinted from: *Water* **2022**, *14*, 256, doi:10.3390/w14020256 . . . . **81**

Spyros Beltaos and Brian C. Burrell
Effects of River-Ice Breakup on Sediment Transport and Implications to Stream Environments: A Review
Reprinted from: *Water* **2021**, *13*, 2541, doi:10.3390/w13182541 . . . . **111**

About the Editor

Bommanna Krishnappan

Dr. Bommanna Krishnappan is retired research scientist from the National Research Institute of Environment Canada in Burlington, Ontario, Canada. His research interests include cohesive sediment transport in rivers, Mathematical modelling of river flows and diffusion and dispersion processes in river flows.

Preface to "Modelling of River Flows, Sediment and Contaminants Transport"

This book is a printed edition of the Special Issue: Modelling of River flows, Sediment and Contaminants Transport that was published in the *Water Journal* under the section: Water Erosion and Sediment Transport. It presents a collection of five papers covering a range of topics, such as wild fire impacts, reservoir sedimentation, basin scale sediment balance, cohesive sediment transport models, and sediment erosion due to ice jams in cold region rivers. The models presented in these papers can serve as useful tools for environmental impact assessment studies in river basins subjected to major land disturbances either due to economic developments in the basin or due to wild fires and other natural disturbances, which are becoming more prevalent recently due to changes in climate.

As a guest editor for this Special Issue, I am grateful to all the contributing authors of these high-quality papers. I am also grateful to *Water Journal* for providing me the opportunity to be part of this fruitful project.

Bommanna Krishnappan
Editor

Editorial

Modelling of River Flows, Sediment and Contaminants Transport

Bommanna G. Krishnappan

Krishnappan Environmental Consultancy, Hamilton, ON L9C 2L3, Canada; krishnappan@sympatico.ca

Citation: Krishnappan, B.G. Modelling of River Flows, Sediment and Contaminants Transport. *Water* 2022, 14, 649. https://doi.org/ 10.3390/w14040649

Received: 11 February 2022
Accepted: 15 February 2022
Published: 19 February 2022

Publisher's Note: MDPI stays neutral with regard to jurisdictional claims in published maps and institutional affil-iations.

Economic development projects in river basins, involving mining, forestry, agriculture and urban developments, invariably impact the aquatic ecosystems of the basin. To ensure that these projects are sustainable, environmental impact assessments need to be carried out. These assessments in turn require a thorough understanding of river dynamics, sediment production and transport and sediment–contaminant interactions and transport in rivers. The modelling of river flows, sediment and contaminant transport can serve as useful tools for carrying out these assessments. For this special issue, papers addressing these aspects were sought, with special emphasis on river morphology, cohesive sediment transport processes and sediment contaminant interactions. All together five high quality papers were received in response to our request. A summary of all five papers is given here to highlight their contributions to the field.

The first paper [1] is an article by Stone, Krishnappan, Silins, Emelko, Williams, Collins and Spencer. It proposes a new modelling strategy, which integrates four existing numerical models (MOBED, RIVFLOC, RMA2 and RMA4) to model cohesive sediment transport in the Oldman River watershed, in Alberta, and to route the sediment from the upland sources, through three rivers in the plains to a downstream reservoir. The MOBED model is a coarse-grained, non-cohesive sediment transport model and it also calculates the unsteady and non-uniform river flows in one dimension. The RIVFLOC model is a cohesive sediment transport model, which calculates the flocculation of cohesive sediments explicitly in two-dimensional river flows. The RMA2 and RMA4 are reservoir models that calculate the depth averaged velocity components in the horizontal plane and the dispersion and settling of fine-grained sediment in reservoirs. The input data for these models were generated by field measurements carried out in the rivers and reservoir. The models were calibrated and applied using data from the long-term monitoring programs, established in the reference (unburned), and fire impacted watersheds. The model predictions in rivers show that the deposition of fine sediment to the bed occurs not because of the traditional deposition process, but because of the entrapment (ingress of fine material into the coarse substrate) process. Predictions in the reservoir show that the depositional characteristics of burned sediment differ from that of unburned sediment. The predicted depositional patterns show a spatial variation within the reservoir, which can lead to zones of increased internal loading of phosphorus to reservoir water columns, thereby increasing the potential for algae proliferation. Because of the growing threats to water resources due to wildfires, the modelling strategy proposed in this paper can be used to model the propagation of fine sediment and the associated nutrients and contaminants to reservoirs, under different flow conditions and land use scenarios. Therefore, the modelling framework is a valuable tool for water resources management and watershed planning.

The second paper [2] is an article by Jia, Zhou, Shao and Zhang, describing a three-dimensional numerical simulation of fine-grained sediment deposition in a large reservoir in China (Three Gorges Dam Reservoir on the Yangtze River). They employed an existing three-dimensional model of water flow, based on Reynolds' equations, a three-dimensional mass balance equation for sediment transport and deposition and a two-dimensional gravity-driven model of fluid mud, formulated by the authors. The three-dimensional model predicts the sediment deposition in the reservoir, based on flow patterns, sediment dispersion and settling, and a mass balance equation calculates the bed level changes during deposition. As the fine sediment deposits on the reservoir bed, the sediment accumulates and forms a fluid mud, with high water content. The fluid mud then flows laterally because of the lateral slope of the bed of the reservoir until it reaches the thalweg (deepest part of the reservoir). In this paper, the authors have formulated a two-dimensional model to calculate the movement of the fluid mud and to redistribute the deposited sediment. Using field measurements of thickness of sediment deposits in the Three Gorges Reservoir, the authors have shown that the computed thickness of the sediment deposits agree well with the measured data. The approach proposed by the authors has the potential to be applicable to sediment deposition in large reservoirs, where the formation of a fluid mud layer is a possible scenario.

The third paper [3] is an article by Datok, Sauvage, Fabre, Laraque, Ouillon, Moukandi N'kaya, Sanchez-Perez, describing the sediment balance estimation of the 'Cuvette Centrale' of the Congo River basin, using the SWAT hydrological model. The Congo River basin is the second largest continuous tropical rain forest in the world and houses the largest peatland complex in Africa. The peatland complex is encircled by the 'Cuvette Centrale', which is a central depression in the heart of the basin. The Soil and Water Assessment Tool (SWAT) is a physically based hydrological model, capable of integrating environmental data, such as climate, soil, land-cover and topographical features, and simulating the hydrological processes, such as surface runoff, infiltration, evapotranspiration, lateral flow, percolation to shallow and deep aquifers and channel routing. The model was run for the period 2000–2012 and calibrated using the measured data from the main gauging station of the basin. The calibrated model was used to estimate the sediment balance for the 'Cuvette Centrale' and found that it acts like a big sink for sediment, trapping up to 23 megatons of sediment produced upstream every year.

The fourth paper [4] is a review article by Krishnappan. It provides a review of a semi-empirical modelling approach for cohesive sediment transport in river systems. Because of the large number of controlling parameters for the transport processes of cohesive sediment, cohesive sediment transport models are invariably semi-empirical, and the model parameters are often measured using a rotating circular flume in the laboratory. One such model, called RIVFLOC, developed by Krishnappan, is reviewed in this paper. The parameters that need to be determined for the application of the model include the critical shear stress for erosion, critical shear stress for deposition, according to the definition of Partheniades, critical shear stress for deposition, according to the definition of Krone, cohesion parameter governing the flocculation of the cohesive sediment and a set of empirical parameters that define the density of the flocs, in terms of the size of the flocs. Measurement of these parameters, using a rotating circular flume, located in the National Water Research Institute in Burlington, ON, Canada, is highlighted in the paper. Application of the RIVFLOC model to different river systems was examined and the resulting parameters for different river systems were compared. The paper concludes that the entrapment process of fine sediment (ingress of fine material into the coarse bed sediment) is an important process that needs to be taken into account when dealing with the cohesive sediment transport in rivers.

The fifth and final paper [5] is by Beltaos and Burrell. This is a review paper and it deals with the effects of river-ice breakup on sediment mobilization and transport. In cold regions of the world, where rivers can freeze during winter months and can flow under intact or broken pieces of ice cover, flows under a solid ice cover can have lower potential for sediment transport, because of the increased boundary resistance and reduced flow velocities. During the breakup of ice covers, on the other hand, a greater potential for the erosion of sediment occurs, due to rising discharge and moving ice and highly dynamic waves that form upon the ice jam release. Under such conditions, sharp increases in suspended sediment concentrations can occur. In this paper, the authors review the basics of river sediment erosion and transport of relevant phenomena that occur during the breakup of river ice, and the datasets of measured and inferred suspended sediment concentrations in different rivers. Possible effects of river characteristics on seasonal sediment supply and the implications of increased sediment supply on water quality and ecosystem functionality were also discussed. The modelling of sediment transport, under the highly dynamic flow conditions that prevail upon and after release of ice jams, by coupling a non-hydrostatic three-dimensional hydrodynamic model, with bed erosion and deposition equations to track the changing bathymetry of the riverbed, is referenced.

Funding: This research received no external funding.

Institutional Review Board Statement: The study was conducted in accordance with the Declaration of Helsinki.

Acknowledgments: The author is grateful to all the contributors to this special issue.

Conflicts of Interest: The author declares no conflict of interest.

References

1. Stone, M.; Krishnappan, B.G.; Silins, U.; Emelko, M.B.; Williams, C.H.S.; Collins, A.L.; Spencer, S.A. A New Framework for Modelling Fine Sediment Transport in Rivers Includes Flocculation to Inform Reservoir Management in Wildfire Impacted Watersheds. *Water* **2021**, *13*, 2319. [CrossRef]
2. Jia, D.; Zhou, J.; Shao, X.; Zhang, X. 3D Numerical Simulation of Gravity-Driven Motion of Fine-grained Sediment Deposits in Large Reservoirs. *Water* **2021**, *13*, 1868. [CrossRef]
3. Datok, P.; Sauvage, S.; Fabre, C.; Laraque, A.; Ouillon, S.; Moukandi N'kaya, G.; Sanchez-Perez, J.-M. Sediment Balance Estimation of the 'Cuvette Centrale' of the Congo River Basin Using the SWAT Hydrological Model. *Water* **2021**, *13*, 1388. [CrossRef]
4. Krishnappan, B.G. Review of a Semi-Empirical Modelling Approach for Cohesive Sediment Transport in River Systems. *Water* **2022**, *14*, 256. [CrossRef]
5. Beltaos, S.; Burrell, B.C. Effects of River-Ice Breakup on Sediment Transport and Implications to Stream Environments: A Review. *Water* **2021**, *13*, 2541. [CrossRef]

Article

A New Framework for Modelling Fine Sediment Transport in Rivers Includes Flocculation to Inform Reservoir Management in Wildfire Impacted Watersheds

Micheal Stone [1], Bommanna G. Krishnappan [2,*], Uldis Silins [3], Monica B. Emelko [4], Chris H. S. Williams [3], Adrian L. Collins [5] and Sheena A. Spencer [6]

[1] Environmental Studies, University of Waterloo, Waterloo, ON N2L 3G1, Canada; mstone@uwaterloo.ca
[2] Environment Canada, Burlington, ON L7R 4A6, Canada
[3] Department of Renewable Resources, University of Alberta, Edmonton, AB T6G 2G7, Canada; usilins@ualberta.ca (U.S.); chw1@ualberta.ca (C.H.S.W.)
[4] Department of Civil and Environmental Engineering, University of Waterloo, Waterloo, ON N2L 3G1, Canada; mbemelko@uwaterloo.ca
[5] Sustainable Agriculture Sciences Department, Rothamsted Research, North Wyke, Okehampton EX20 2SB, UK; Adrian.collins@rothamsted.ac.uk
[6] Ministry of Forests, Lands, Natural Resource Operations and Rural Development, Government of British Columbia, Penticton, BC V2A 7C8, Canada; sheena.spencer@gov.bc.ca
* Correspondence: krishnappan@sympatico.ca

Citation: Stone, M.; Krishnappan, B.G.; Silins, U.; Emelko, M.B.; Williams, C.H.S., Collins, A.L., Spencer, S.A. A New Framework for Modelling Fine Sediment Transport in Rivers Includes Flocculation to Inform Reservoir Management in Wildfire Impacted Watersheds. *Water* **2021**, *13*, 2319. https://doi.org/10.3390/w13172319

Academic Editor: Maria Mimikou

Received: 31 July 2021
Accepted: 20 August 2021
Published: 24 August 2021

Publisher's Note: MDPI stays neutral with regard to jurisdictional claims in published maps and institutional affiliations.

Abstract: Fine-grained cohesive sediment is the primary vector for nutrient and contaminant redistribution through aquatic systems and is a critical indicator of land disturbance. A critical limitation of most existing sediment transport models is that they assume that the transport characteristics of fine sediment can be described using the same approaches that are used for coarse-grained non-cohesive sediment, thereby ignoring the tendency of fine sediment to flocculate. Here, a modelling framework to simulate flow and fine sediment transport in the Crowsnest River, the Castle River, the Oldman River and the Oldman Reservoir after the 2003 Lost Creek wildfire in Alberta, Canada was developed and validated. It is the first to include explicit description of fine sediment deposition/erosion processes as a function of bed shear stress and the flocculation process. This framework integrates four existing numerical models: MOBED, RIVFLOC, RMA2 and RMA4 using river geometry, flow, fine suspended sediment characteristics and bathymetry data. Sediment concentration and particle size distributions computed by RIVFLOC were used as the upstream boundary condition for the reservoir dispersion model RMA4. The predicted particle size distributions and mass of fine river sediment deposited within various sections of the reservoir indicate that most of the fine sediment generated by the upstream disturbance deposits in the reservoir. Deposition patterns of sediment from wildfire-impacted landscapes were different than those from unburned landscapes because of differences in settling behaviour. These differences may lead to zones of relatively increased internal loading of phosphorus to reservoir water columns, thereby increasing the potential for algae proliferation. In light of the growing threats to water resources globally from wildfire, the generic framework described herein can be used to model propagation of fine river sediment and associated nutrients or contaminants to reservoirs under different flow conditions and land use scenarios. The framework is thereby a valuable tool to support decision making for water resources management and catchment planning.

Keywords: cohesive sediment; erosion; water supply; turbidity; gravel bed river; ingress; watershed management; source water protection; climate change adaptation; landscape disturbance

1. Introduction

Forested regions provide approximately 86% of surface water supplies in the United States (Caldwell et al. [1]) and more than 58% for the largest Canadian urban and rural communities as well as the majority of Canadian Indigenous communities. Rivers draining forested landscapes are important for the provision of high-quality source water and support of healthy aquatic ecosystems. While various anthropogenic disturbances (e.g., harvesting, recreational use, land clearing for agriculture or resource extraction) in these critical water-bearing landscapes can alter erosion and runoff to, and subsequent sedimentation within, receiving waters (Kastridis and Kamperidou [2], Vacca et al. [3]), the increasing frequency and severity of landscape disturbance by wildfire has raised urgent concerns about degraded and more variable water quality and its implications for the provision of safe drinking water (Vörösmarty [4]; Emelko et al. [5]; International Panel on Climate Change (IPCC) [6]). Accordingly, watershed management is directly linked to national security in some regions (Caldwell et al. [1]). National and international commitment to source water protection in forested watersheds has been increasingly advocated (Vörösmarty et al. [7]; Emelko and Sham [8]).

Wildfire is the most severe large-scale landscape disturbance in critical forested source water regions (Emelko and Sham [8]; Vose et al. [9]; Khan et al. [10]). Recent increases in the size and severity of wildfires related to climate warming (Westerling et al. [11]; Flannigan et al. [12]) have been shown to degrade terrestrial ecosystems, ecological processes and functions, and surface water quality (Benda et al. [13]; Khan et al. [10]), whilst threatening human life and property (Kinoshita et al. [14]).

Wildfire impacts on water quality vary because of differences in physiographic setting as well as hydro-climatic and landscape factors such as wind speed, moisture conditions, and vegetation type (Bisson et al. [15]; Vörösmarty et al. [7]; Silins et al. [16]; Lucas-Borja et al. [17]; Plaza-Alvarez et al. [18]). The effects of severe wildfire on water quality have been observed at the watershed scale (Emmerton et al. [19]; Silins et al. [20]; Emelko et al. [21]) wherein increases in sediment yields and instream sedimentation have been measured (Benda et al. [13]; Tobergte and Curtis [22]). Wildfire can exacerbate the impact of extreme rain events, which can mobilise and transport significant amounts of sediment (López-Vicente et al. [23], Malmon et al. [24]). Critically, in regions underlain by glacial deposits, such as the eastern slopes of the Rocky mountains in Alberta, Canada, fine sediment delivery from terrestrial to aquatic systems can be elevated and prolonged (Silins et al. [25]; Stone et al. [26]); it also contributes to the downstream transfer and fate of sediment-associated phosphorus which, in turn, influences reservoir water quality (Stone et al. [26]; Silins et al. [20]; Emelko et al. [21]).

Sediment is the primary vector for nutrient and contaminant redistribution through aquatic systems (Horowitz and Elrick [27]; Ongley et al. [28]; Chapman et al. [29]) and is a critical indicator of land disturbance (Walling and Collins [30]). Excessive amounts of fine sediment can reduce light transmission in high quality streams, decrease flow through interstitial gravels and lower oxygen supply in spawning habitat (Collins et al. [31]; Wood and Armitage [32]). Accurate representation of fine sediment transport informs the fate and bioaccumulation of many toxic substances and the availability of limiting nutrients such as phosphorus, which contribute to eutrophication in aquatic systems. Accordingly, there is an important need to model fine sediment transport in aquatic systems as robustly as possible.

A critical limitation of most existing sediment transport models is that they assume that the transport characteristics of fine-grained cohesive sediment can be described using the same approaches that are used for coarse-grained non-cohesive sediment, thereby ignoring the fundamental tendency of fine sediment to flocculate (Partheniades [33]; Krone [34]; Mehta [35]; Lick [36]). Flocculation strongly influences the transport properties (porosity, density, settling velocity) and fate of fine sediment and associated contaminants (Lau and Krishnappan [37], Krishnappan [38,39]). Thus, while coarse-grained sediments undergo simultaneous erosion and deposition during transport at constant bed shear stress in aquatic systems, the simultaneous erosion and deposition of fine sediments is not possible and they undergo either deposition or erosion at certain bed shear stresses, but not both (Partheniades and Kennedy [40]; Mehta and Partheniades [41]; Lau and Krishnappan [37];

Krishnappan [38]). If simultaneous erosion and deposition are assumed in transport modelling, fine sediment and associated contaminant concentrations will be under-predicted; in contrast, appropriate representation of mutually exclusive erosion and deposition will preserve the comparatively high concentrations of sediment and associated contaminants that will transport over relatively longer distances from the source (Krishnappan [38,39]). Improved understanding and representation of the mobilisation, flocculation and transport dynamics of fine sediment in watersheds is therefore an essential pre-requisite for land managers seeking to evaluate the fate and impacts of diffuse source pollution resulting from both anthropogenic and natural landscape disturbances (Walling and Collins [30]; Emelko et al. [5]).

To date, modelling efforts to describe disturbance impacts on water have been largely limited to simulating hydrologic and erosion responses (e.g., streamflow, infiltration, runoff). For example, the Soil and Water Assessment (SWAT) model has been applied to compute change in streamflows after wildfire (Rodrigues et al. [42]). Several physically-based transport models have been developed to describe coarse-grained sediment erosion in wildfire impacted landscapes at the plot and hillslope scales (Moody et al. [43]) and landscape evolution models have been applied to simulate gullying, landslides and debris flows in small watersheds (Lancaster et al. [44]; Istanbulluoglu et al. [45]). Frameworks have also been developed to simulate sediment delivery processes from runoff-generated debris flows to reservoirs at the watershed scale; however, they are not data driven and only involve empirical calculation of debris flow volumes and expert-based assumptions regarding fine sediment delivery (Laghans et al. [46]). Post-fire water quality has also been simulated using the MIKE Hydro Basin water quality model with ECOLab (Santos et al. [47]). The MIKE ECOLab module conducts simplified water quality simulations solving generic ordinary differential equations based on user inputs (rather than hard-coded calculations) (Danish Hydraulic Institute [48]). While the MIKE Hydro Basin module does not include either cohesive (i.e., fine-grained) or non-cohesive sediment, the MIKE Hydro River module includes a 1D computational platform for modelling sediment transport. Cohesive sediments are treated as suspended load—cohesive and non-cohesive sediments are transported in the same manner, but cohesive sediments use different erosion and deposition functions (Danish Hydraulic Institute [49]). Critically, this platform requires user definition of the fraction of total load that is suspended (i.e., fine-grained) and does not include flocculation (Danish Hydraulic Institute [49]). Thus, other investigations of climate and land use change impacts on sediment transport that have coupled hydrologic models like SWAT with the MIKE platform (Anand et al. [50]) also suffer from the same important limitation of disregarding the flocculation process.

From a management perspective, there is a critical need to develop and test process-based, as opposed to risk-based, models that simulate fine sediment transport dynamics in rivers and downstream receptors for a range of land disturbance types (Walling et al. [51]; Walling and Collins [30]; Daniel et al. [52]). This need arises from the capacity of process understanding to help inform targeted intervention above and beyond the spatial information generated by risk-based approaches. Given the significance of fine sediment for contaminant transport (Horowitz and Elrick [27]; Ongley et al. [28]) and the influence of flocculation (Lau and Krishnappan [37], Krishnappan [38,39]) on its redistribution and fate in aquatic systems, it is necessary to advance sediment transport models to incorporate flocculation processes explicitly (Summer and Walling [53]). At present, models that specifically quantify fine sediment transport processes and flocculation in river systems at large basin scales are scant. To address this research gap, the objectives of this work were to: (1) formulate a framework that includes explicit description of deposition/erosion processes as a function of bed shear stress and the flocculation process to model fine sediment transport in rivers to inform reservoir management in wildfire impacted watersheds, and; (2) demonstrate the utility of this framework to quantify sediment fluxes to reservoirs and inform post-fire reservoir management. The modelling framework reported herein is applied to three rivers in the Oldman watershed located immediately upstream of the Oldman Reservoir. Detailed hydrometric and sediment monitoring surveys were conducted in the upper part of the watershed to calibrate flow and fine sediment transport models. Long term hydrometric and sediment data from reference (unburned) and wildfire impacted tributaries of the Crowsnest River are used to demonstrate the utility of the framework to quantify sediment transport and depositional fluxes in the river and reservoir.

2. Materials and Methods

2.1. Study Area

The study was conducted in the upper Oldman River watershed situated along the front range of the Rocky Mountains in southwest Alberta, Canada (Figure 1). The three main rivers that drain the watershed are the Oldman River from the north (area = 1923 km^2, mean elevation = 1741 m.a.s.l (3093–1110), mean basin slope = 44% (928–0%)), the Crowsnest River from the west (area = 1006 km^2, mean elevation = 1554 m.a.s.l (2803–1110), mean basin slope = 40% (611–0%)) and the Castle River from southwest (area = 1224 km^2, mean elevation = 1623 m.a.s.l (2737–1076), mean basin slope = 47% (891–0%)). All three rivers discharge into the Oldman Reservoir, which was created in 1993 when the Oldman Dam was built as an instream storage facility (~500 million m^3) for irrigation, power generation and recreational activities. The reservoir is located about 30 km downstream of the town of Blairmore in the Crowsnest Pass, where there was a severe forest fire (Lost Creek Wildfire) in 2003. The reservoir is ~20 km in length, with a maximum depth of ~65 m and a maximum width of ~2 km.

Mean annual precipitation and air temperature (town of Coleman, Alberta in the central study region) is 582 mm/yr and 3.6 °C, respectively with mean summer (July) and winter (December) air temperatures of 14.3 and −7.4 °C, respectively. Land cover in the study region spans high elevation alpine (exposed sedimentary bedrock and alpine shrubs), sub-alpine forests (dominated by *Abies lasiocarpa* and *Picea engelmannii*) and lower elevation montane forests (dominated by *Pinus contorta var. latifolia* and *Pseudotsuga menziesii var. glauca*) in western and central regions, with mixed native rangeland vegetation the lower elevation eastern region.

In 2003, the Lost Creek fire severely burned a near contiguous area of 21,065 ha on the eastern slopes of the Rocky Mountains in southern Alberta (Silins et al. [25]) which is one of the highest source-water yielding regions in the province (Figure 1). The subsequent post-fire hydrological changes dramatically increased sediment production (Silins et al. [25]) and accelerated the propagation of cohesive sediment and associated contaminants to downstream environments (Stone et al. [26]; Emelko et al. [21]) including the Oldman Reservoir. The new modelling framework described herein is applied to route fine-grained (<62.5 μm) sediments from the upper watershed to the reservoir to simulate the response to the wildfire disturbance.

2.2. Modelling Framework

The modelling framework shown schematically in Figure 2 incorporates four existing models: a river flow model (MOBED) that predicts the unsteady and non-uniform flows in rivers under mobile boundary conditions (Krishnappan [54–56]); a fine sediment transport and dispersion model (RIVFLOC) that calculates the dispersion and flocculation of fine-grained sediment in rivers (Krishnappan [57]); a reservoir flow model (RMA2) that simulates two dimensional flow fields (Donnell et al. [58]), and; a water quality model (RMA4) that can compute the dispersion and transport of fine sediment within a reservoir (Letter et al. [59]). The reservoir models RMA2 and RMA4 are part of the TABS-MD modelling system which is linked to an SMS user interface developed by AQUAVEO (www.aquaveo.com (accessed on 21 August 2021)). Specific components of the framework are briefly described below.

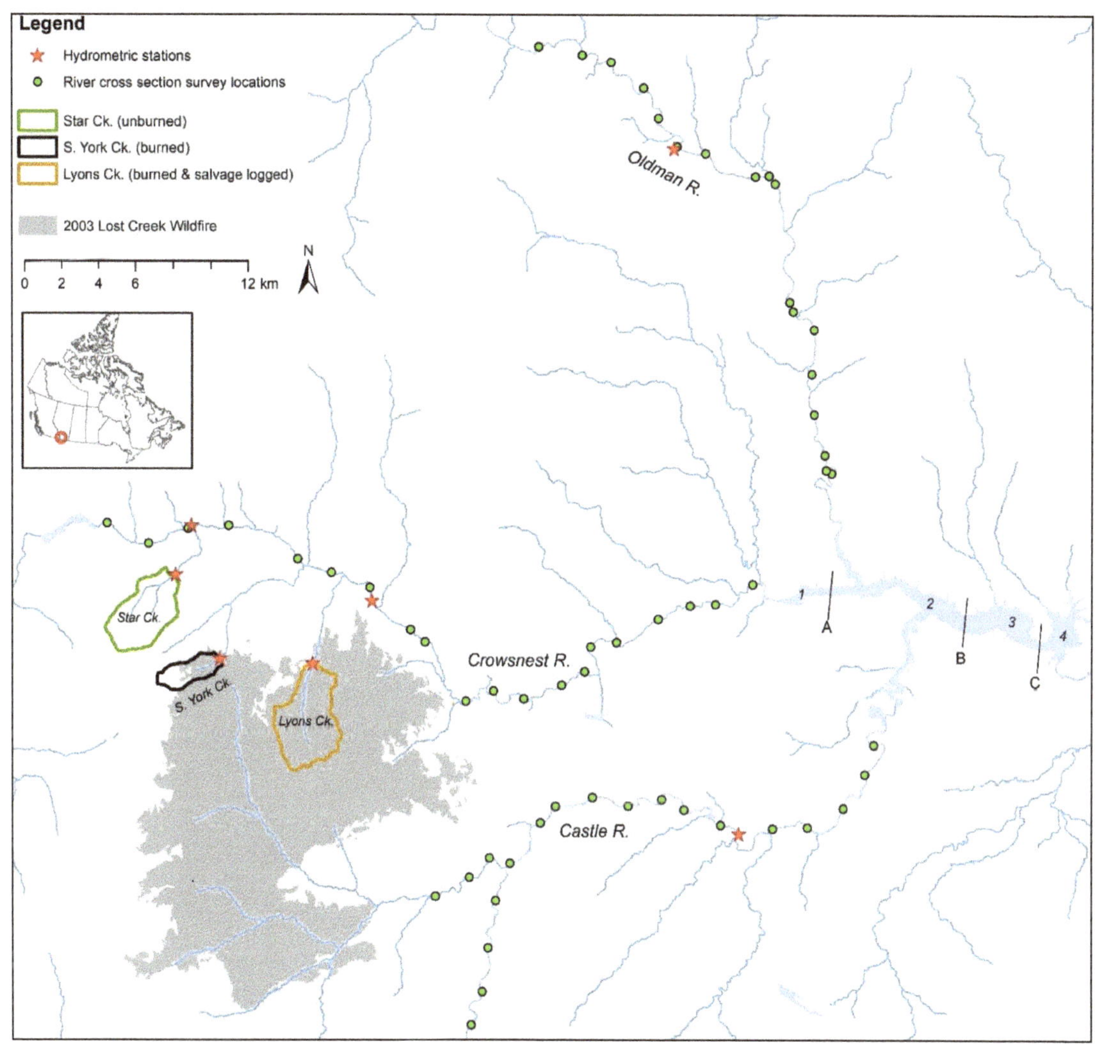

Figure 1. Study area. Upper Oldman River basin including the upper Oldman, Crowsnest and Castle Rivers, locations of cross section surveys, Water Survey of Canada and Southern Rockies Watershed Project hydrometric stations and three gauged unburned, burned, and post-fire salvage logged catchment tributaries to the main stem of the Crowsnest River. Inset shows study region in southwest Alberta, Canada.

2.2.1. Mobile Boundary Flow Model—MOBED

MOBED is an unsteady, mobile boundary flow model based on the numerical solution of St. Venant's equations and a sediment continuity equation (Krishnappan [54]). MOBED calculates flow rate, flow depth and bed shear stress as a function of time and distance in alluvial channels for a given upstream boundary flow condition. The model calculates the transport of coarse-grained bed sediment and resulting changes in the bed level for a full range of feasible flow conditions. MOBED uses a generalized friction factor equation that can be applied to various river types and bed forms (Krishnappan [54–56]). Specific details of the model can be found in Krishnappan [54–56].

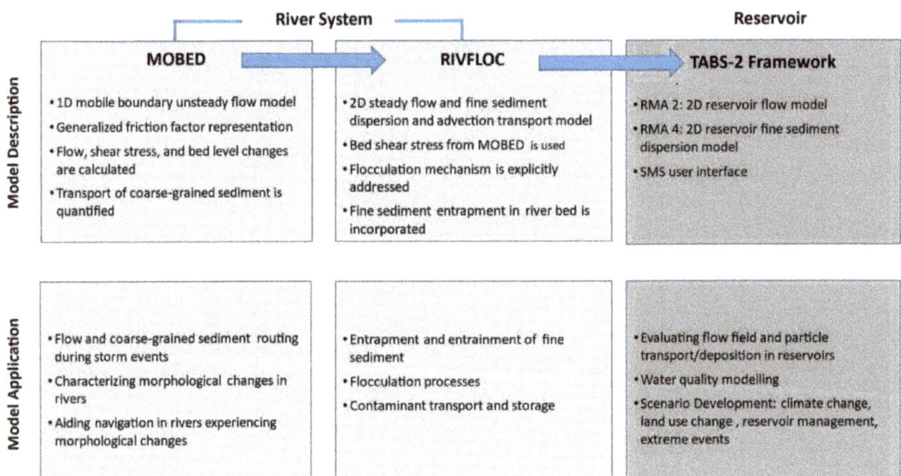

Figure 2. Schematic of the modelling framework.

2.2.2. Fine Sediment Transport Model—RIVFLOC

RIVFLOC is a cohesive sediment transport model which explicitly describes flocculation of fine sediment in the water column (Krishnappan [57]). It consists of two modules: a transport module and a flocculation module. The transport module is based on an advection-diffusion equation expressed in a curvilinear co-ordinate system to treat the transport and mixing of fine sediment entering a river as a steady source. The flocculation module incorporates a coagulation equation, which expresses the mass balance of the sediment during flocculation. Collision mechanisms such as Brownian motion, turbulent fluid shear, inertia of particles and differential settling are considered. The model uses an empirical relationship to express floc density as a function of floc size and a modified Stokes equation to calculate the settling velocity of solids in suspension. Specific details of the model and its application can be found in Krishnappan and Marsalek [60] and Droppo and Krishnappan [61].

2.2.3. TABS-MD with SMS User Interface

TABS-MD with SMS user interface consists of two models: RMA2 is a two-dimensional depth-averaged finite-element hydrodynamic model and RMA4 is a two-dimensional depth-averaged finite-element sediment transport and water quality model. Details of both models are reported in Donnell et al. [58] and Letter et al. [59], respectively. RMA2 and RMA4 are fully integrated by a user interface called SMS which is a graphical pre-and post-processor for numerical surface water models to allow interactive editing and display of finite element networks. Display controls allow the user to adjust color and line contouring to display either bed elevations or model results such as velocity fields and water surface elevations. SMS consists of a data module that includes tools for performing data analysis and interpretation. A mapping module allows the user to create a conceptual model and use background images to interface with the finite element mesh of the computational domain. A mesh module allows the creation of finite element meshes for different hydrodynamic modelling systems. SMS version 11 was used in this study.

2.2.4. Input Data Requirements for the Modelling Framework

The input data requirements for the component models of the modelling framework are shown in Table 1.

Table 1. Model input data requirements and corresponding sources.

Component Models	Data Requirements	Data Sources
MOBED model	Hydraulic geometry and surface water elevation at 2 km intervals along the study reach for each river	2011 cross sectional surveys described in Section 2.2.5
	Bed material size data	
	Flow rate	Water survey of Canada Hydrometric stations for Crownsnest River @ Frank Stn 05AA008; for Castle River @ Ranger Station Stn 05AA028; and for Oldman River @ Range Road Stn 05AA035
	Frictional parameters	Calibration of MOBED model described in Section 2.2.6.
RIVFLOC model	River geometry data: cross sectional shapes at a number of sections along the river	2011 Cross sectional survey described in Section 2.2.5
	Particle size distribution at the upstream boundary of the modelling domain	2015 survey in the upper Crowsnest River (LISST measurements) described in Section 2.2.7.
	Suspended sediment concentration at the upstream boundary of the modelling domain	2015 survey in the upper Crowsnest River described in Section 2.2.7
	Relationship between the floc size and floc density	2015 survey in the upper Crowsnest River described in Section 2.2.7.
	Bed shear stress distribution in the modelling domain	Provided by the MOBED model predictions
	Critical shear stress for deposition of fine sediment	Based on erosion and deposition experiments in annular flume Stone et al. [62]
	Cohesion parameter, β	Calibration parameter for RIVFLOC model-described in Section 2.2.8
RMA2 model	Bathymetry data to formulate the finite element mesh	Provided by existing reservoir bathymetric data
	Flow rate at the upstream boundary of the reservoir	Provided by the output of the RIVFLOC model
RMA4 model	Two dimensional lateral velocity distribution in the reservoir	Provided by the output of the RMA2 model
	Suspended sediment concentration at the upstream boundary of the reservoir	Provided by the output of the RIVFLOC model
	Size distribution of the suspended sediment entering the reservoir at the upstream boundary of the reservoir	Provided by the output of the RIVFLOC model

2.2.5. Cross-Section Survey of the Crowsnest, Castle and Oldman Rivers

In 2011, cross-section elevation and river-bed cobble size surveys were conducted at ~2 km intervals along 54, 46 and 46 km reaches of the Crowsnest, Castle and Oldman Rivers, respectively (Figure 1). Hydrometric data collected at Water Survey of Canada gauging stations (Crowsnest River, 05AA008; Castle River, 05AA028; Oldman River, 05AA035) were used to set the boundary conditions for the models. The physical characteristics (slope, width and bed materials) of the Castle, Crowsnest and Oldman study reaches are variable, and the river-bed elevation decreased by ~250 m over a distance of ~50 km. The average slope of the Crowsnest, Castle, and Oldman Rivers is 0.44%, 0.43% and 0.46%, respectively. The Crowsnest River is narrower (10 to 20 m wide) than the Castle and Oldman Rivers which can vary in width from 70 to 100 m. Bed materials in these rivers consist primarily of coarse sand, pebbles and cobbles. River-beds are armoured and stable during low flows but mobilisation and transport of bed material can occur during flood stages. Photographs of the bed materials of the three river sections are shown in Figure 3.

Figure 3. Sample photographs of river-beds: (**a**) Crowsnest River, (**b**) Castle River and (**c**) Oldman River.

2.2.6. Calibration of MOBED

Using the cross-sectional shape and the river profile data, the MOBED model was calibrated for the flow conditions that existed during the cross-section shape surveys. At the upstream boundary, a flow rate boundary condition was used and, at the downstream boundary, a stage–discharge relationship derived from the measured data was used. Bed material size at each cross-section was estimated from underwater photographs (using a metal frame grid which measures 25 cm $\times$ 25 cm) of the bed material taken at five equidistant locations across each section. An image analysis system was used to calculate the bed material size distribution characteristics such as D_{35}, D_{50}, and D_{65} needed for the MOBED model. Calibration of the MOBED model was completed by running the model for flow rates that existed when the cross-section surveys were carried out and the calculated water surface elevations were compared with the measured data. A Manning roughness coefficient that produced the best match between the predicted and measured data was determined. Comparisons of the measured and predicted water surface elevations for all three river reaches are shown in Figure 4. A Manning's roughness parameter of 0.070 was estimated for all of the three study reaches and this value is comparable to previously reported values for similar sized bed materials (Hey, et al. [63]). The accuracy of calibration was estimated by comparing the predicted water surface elevation and the measured water surface elevation and the percent deviation varied in the range of ± 0.1 to ± 1.0 percent of the measured water surface elevation. The MOBED model had been field verified by a number of earlier studies (Krishnappan [54–56]) and hence no attempt was made here to validate the model.

The bed shear stress predicted by the MOBED model for the three river reaches is shown in Figure 5. The magnitude of the bed shear stresses ranged from 10 to 50 Pa. The variability of the predicted bed shear stress is due to variation in river geometry (depth and width) governing mean flow characteristics.

2.2.7. Fine Sediment Transport Survey in 2015 in the Upper Crowsnest River

In 2015, a detailed survey of suspended sediment concentrations and effective particle size distributions was conducted in situ for two different flow conditions in the upper Crowsnest River over a reach of about 20 km. Effective particle size distributions were measured using a laser particle size analyzer (LISST 100X; Sequoia Scientific, Bellevue, WA, USA) at several locations near the confluences of three tributary inflows to the Crowsnest River including Star Creek (1035 ha, mean slope = 27%, undisturbed), South York Creek (365 ha, mean slope = 56%, 54% burned) and Lyons Creek (1309 ha, mean slope = 46%, 100% burned and 20% salvage logged). At each confluence, discharge, effective particle size distributions and suspended sediment concentrations were measured in, above and below the confluence of each tributary with the Crowsnest River (see Figure 6).

This measurement campaign provided the input data necessary for setting up the RIVFLOC model (see Table 1). Using these data (flow rate, suspended sediment concentration and the effective particle size distribution of fine sediment at the upstream boundary of the modelling domain), the values of bed shear stresses predicted by MOBED and the critical shear stress for the deposition of the sediment given by the laboratory investigation by Stone et al. [62], the RIVFLOC model was applied to the upper 20 km stretch of the Crowsnest River and the cohesion parameter, β was established as part of the calibration procedure. Details of this calibration procedure are described in Section 2.2.8.

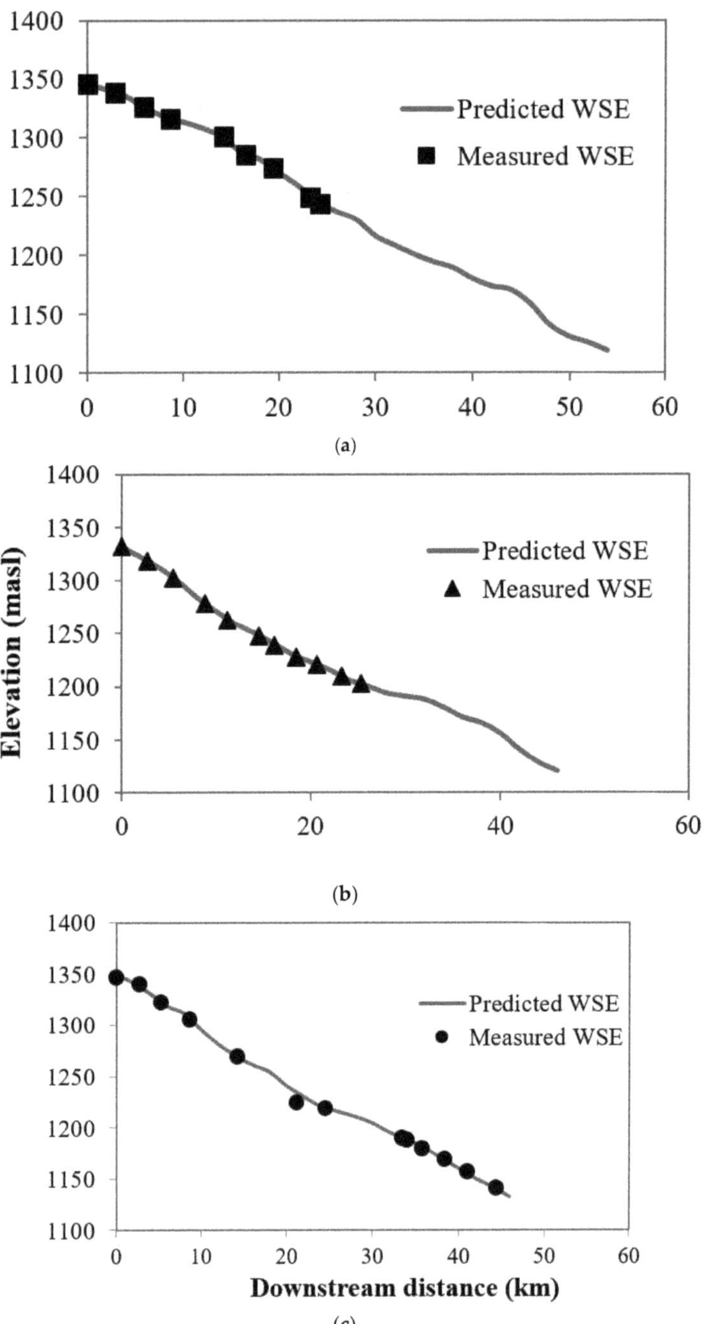

Figure 4. Comparisons between measured and predicted water surface elevations for the Crowsnest River reach ((**a**) Flow rate = 2.65 m^3/s; Manning's n = 0.070), Castle River reach ((**b**) Flow rate = 4.0 m^3/s; Manning's n = 0.070) and Oldman River reach ((**c**) Flow rate = 5.0 m^3/s; Manning's n = 0.070).

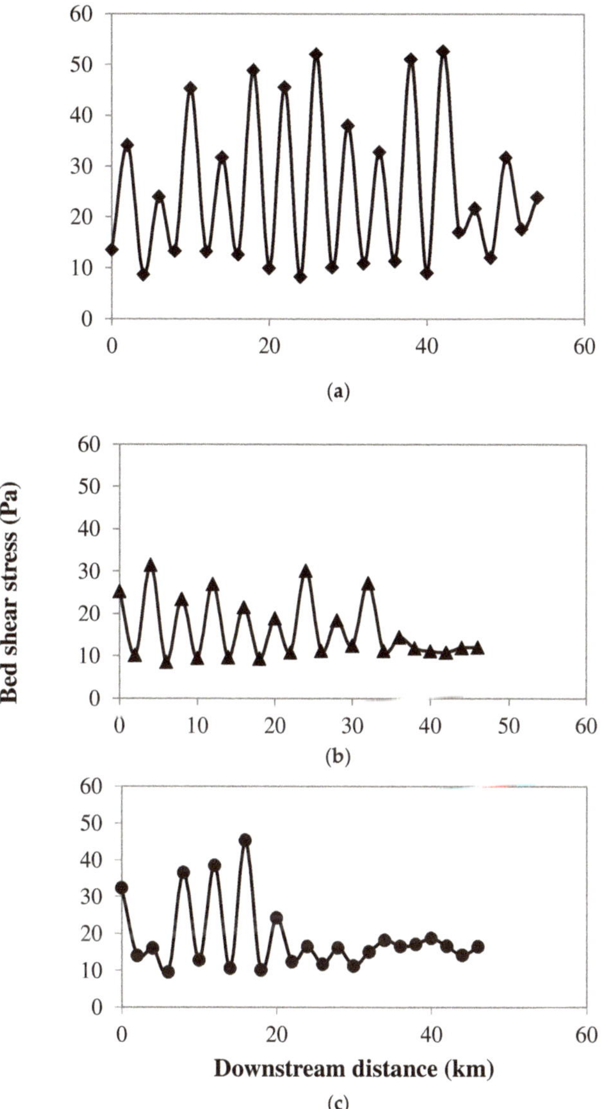

Figure 5. Variation of predicted bed shear stress with distance along the river reaches: Crowsnest River ((**a**) Flow rate = 2.65 m^3/s; Manning's n = 0.070), Castle River ((**b**) Flow rate = 4 m^3/s, Manning's n = 0.070), Oldman River ((**c**) Flow rate = 5 m^3/s, Manning's n = 0.070).

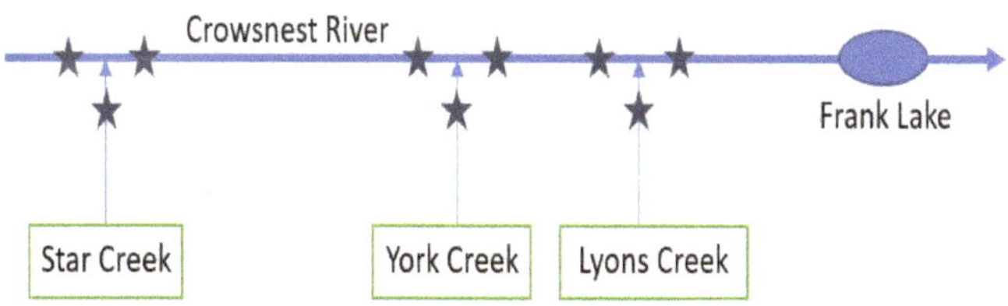

Figure 6. A schematic diagram of the sampling sites in the upper Crowsnest River.

2.2.8. Calibration of RIVFLOC

Hydrometric and sediment survey campaigns were carried out, in May and June 2015. The flow rate and suspended sediment concentration data are summarized in Tables 2 and 3. In May 2015, sediment concentrations were lower in the tributaries than the main stem of the Crowsnest River. Mixing occurred at the confluence of tributary inflows with the main channel and concentrations of suspended sediment decreased due to a dilution effect (Table 2). Calculation of sediment concentrations downstream of the confluence with the modelling framework, assuming complete mixing and no interaction of the sediment with the bed, yielded values that agree favourably with the measured concentrations (see measured and predicted values in Table 2). The contribution of sediment to the main channel from tributary inflow accounted for 1.0 to 3.3% of the sediment mass in the Crowsnest River. In June 2015, downstream trends in suspended sediment were similar to those observed in May, except that flow rates and sediment concentrations were much higher (Table 3). At this time, the contribution of sediment loading from the tributaries ranged from 3 to 8%.

Representative effective particle size distributions of suspended sediment measured directly in the water column for two stations (upstream of Star Creek confluence and downstream of Lyons Creek confluence) in the Crowsnest River are shown in Figure 7. Particle size distributions at these two stations were similar for smaller particle size ranges, but the distributions deviate from each other as particle size increases. For particles > 200 μm, the distribution corresponding to the station above Star Creek was slightly smaller in comparison to the distribution corresponding to the station below Lyons Creek. This deviation is due to flocculation of suspended solids in the water column. Evidence of flocculation is provided by direct observation using sediment collected on a filter paper and observed using an inverted microscope (Figure 8). A representative photomicrograph confirms the presence of flocs in the sediment population.

Table 2. Summary of suspended sediment measurements in the upper Crowsnest River (CNR) and tributaries (May 2015). Study sites relate to those shown in the schematic in Figure 6.

Study Site	Distance (m)	Measured Discharge (m^3/s)	Estimated Discharge (m^3/s)	Measured Concentration (cc/m^3)	Predicted Concentration (cc/m^3)	% Difference	Sed Load (cc/s)	% Contribution
CNR-u/s Star	5000	4.97		23.3			116.4	
Star	5500	0.31		6.0			1.8	1
CNR-d/s Star	6000		5.28	23.0	22.4	2.4	118.2	
CNR-u/s York	14,000	6.9		25.6			176.7	
York	14,500	0.82		7.7			6.3	3.3
CNR-d/s York	15,000		7.72		23.7		183.0	
CNR-u/s Lyons	16,000	7.45		25.2			188.0	
Lyons	16,500	0.51		6.2			3.2	1.7
CNR-d/s Lyons	17,000		7.96	22.8	22.6	0.6	191.2	
Frank Lake	20,000	8.45						

Table 3. Summary of suspended sediment measurements in the upper Crowsnest River (CNR) and tributaries (June 2015). Study sites relate to those shown in the schematic in Figure 6.

Study Site	Downstream Distance in m	Measured Discharge (m³/s)	Estimated Discharge (m³/s)	Measured Concentration (cc/m³)	Predicted Concentration (cc/m³)	% Difference	Sed Load (cc/s)	% Contribution
CNR-u/s Star	5000	13.3		33.1			444	
Star	5500	1.1		32.5			34.9	3.1
CNR-d/s Star	6000		14.4	32.5	33.1	1.8	475	
CNR-u/s York	14,000		20.6	51.7			1066	
York Creek	14,500	2.7		22.8			62.0	5.5
CNR-d/s York	15,000		23.4	46.7	48.3	3.4	1129	
CNR-u/s Lyons	16,000		23.4	44.9			1094	
Lyons	16,500	2.4		35.8			85.9	7.6
CNR-d/s Lyons	17,000		25.8	46.3	44.1	4.8	1135	
Frank Lake	20,000							

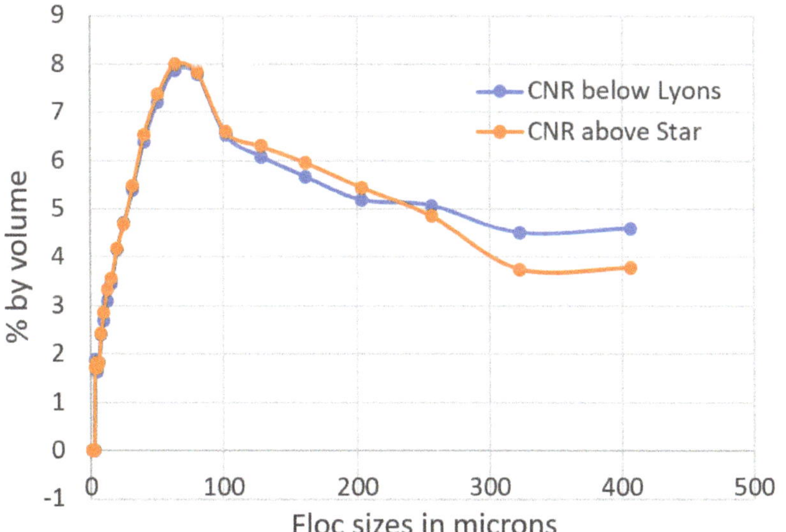

Figure 7. Effective particle size distributions of suspended sediment at two stations in the upper Crowsnest River.

The flocculation module of RIVFLOC requires data on the density and settling velocity of flocs which are dependent on floc size (Lau and Krishnappan [64]). The relationship between floc density and floc size was determined by utilizing simultaneous measurements of volumetric concentration of suspended sediment as measured by the LISST 100X during the 2015 sediment survey and the mass concentration of suspended sediment samples collected during the same survey. The relationship between size, density and settling velocity of flocs is presented in Figure 9.

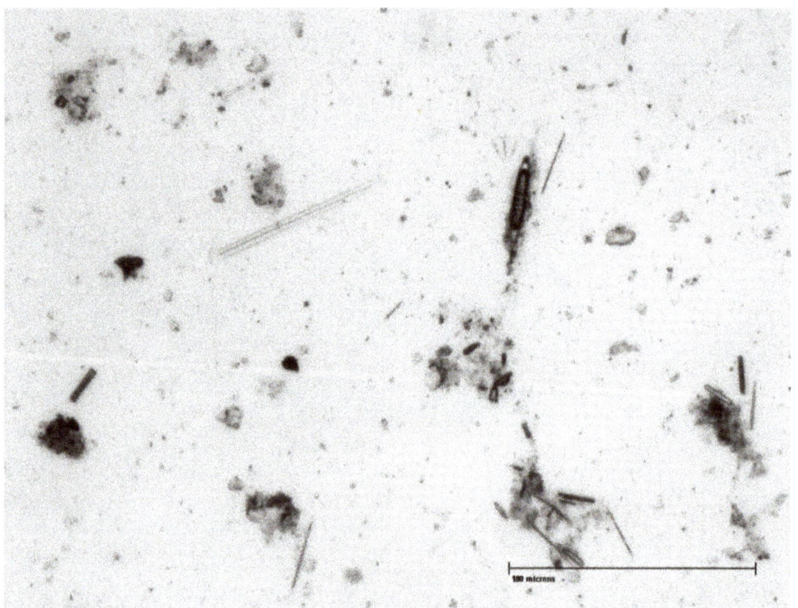

Figure 8. Photomicrograph of the suspended sediment in the Crowsnest River.

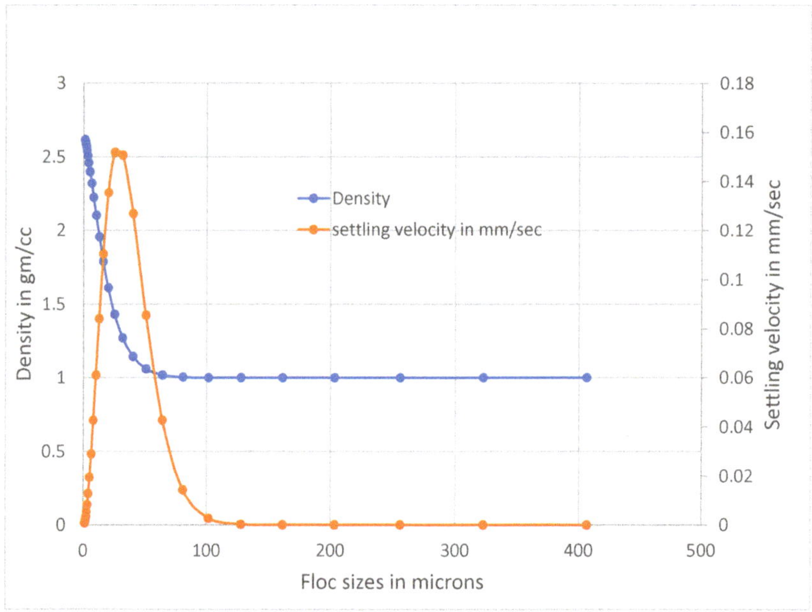

Figure 9. Density and settling velocity of sediment flocs as a function of floc sizes for suspended sediment in the Crowsnest River.

Because water is incorporated into the assemblage of particles in larger flocs, these data show that floc density decreases with increasing floc size (Droppo [65]; Krishnappan and Stephens [66]). The relationship between decreasing density as a function of floc size results in a complex floc settling behaviour. For smaller floc sizes, the settling velocity increases as a function of floc size and reaches a maximum value but with further increases in floc size, the settling velocity decreases and approaches a value close to zero for larger flocs. This settling behaviour of velocity vs floc size has also been observed in previous studies (Droppo [65]; Krishnappan [39]).

In RIVFLOC, the sediment flux (q_{sd}) at the sediment-water interface is specified using the Krone's [34] formulation. According to this formulation, q_{sd} is given by

$$q_{sd} = p w_s C_b \tag{1}$$

where p is a measure of the probability that a floc, settling to the bed, stays in the bed. Krone proposed a relationship for p as:

$$p = \left(1 - \frac{\tau}{\tau_{crd}}\right) \text{ for } \tau < \tau_{crd} \tag{2}$$

$$p = 0 \text{ for } \tau > \tau_{crd} \tag{3}$$

where τ is the bed shear stress and τ_{crd} is the critical shear stress for deposition, which is defined as the bed shear stress above which none of the initially suspended sediment would deposit.

For applying the RIVFLOC model to the upper Crowsnest River reach, the upstream boundary conditions were specified using the distribution of volumetric sediment concentration and size distribution measured at the Crowsnest River station located upstream of the confluence with Star Creek. To specify the boundary condition at the sediment-water interface, the bed shear stress and the critical shear stress for deposition of the sediment are required, as previously described. Bed shear stresses for the flow field were computed using MOBED. The critical shear stresses for the deposition of sediment were obtained from a laboratory flume study carried out by Stone et al. [62]). The measured critical shear stress for erosion (τ_{crit}) of burned and un-burned cohesive sediments from streams in the Castle River watershed (Figure 1) for different consolidation and bio-stabilization conditions (2, 7, 14 days) are presented in Table 4. The critical shear stresses thresholds for deposition of suspended solids in the Crowsnest River were deduced from the values shown in Table 4 using the results from earlier laboratory studies on cohesive sediments from different sources (Krishnappan and Stephens [66]; Krishnappan [38]; Krishnappan et al. [67]). Previous laboratory studies on cohesive sediments demonstrate that the critical shear stress for deposition (i.e., the shear stress below which all of the initially suspended sediment would deposit) is about one half of the value for the critical shear stress for erosion (Partheniades [33]). These studies show that the critical shear stress for deposition, as defined by Partheniades [33]), which is the shear stress below which all of the initially suspended sediment would deposit, is about one tenth of the critical shear stress for deposition as defined by Krone [34], i.e., the shear stress above which none of the sediment in suspension would deposit. Using these two results, the critical shear stresses for deposition as defined by Krone [34]), which are needed for the RIVFLOC model were calculated by multiplying the values in Table 4 by five and the resulting values are listed in Table 5. Since the bed shear stress (τ) predicted by MOBED (10 to 50 Pa) is considerably larger than the critical shear stress for deposition (Table 5), the depositional flux becomes zero and under this boundary condition, RIVFLOC predicts a constant fine sediment concentration as a function of distance along the study reach, suggesting that the fine sediment is simply propagated through the river channel system.

The cohesion parameter, β required for RIVFLOC was determined by applying the model to a sub-reach of the Crowsnest River between the confluences of Star Creek and Lyons Creek. An effective particle size distribution measured at a cross-section above the confluence of Star Creek using a LISST 100X during the 2015 sediment survey was used as the upstream boundary condition. The size distribution of sediment flocs for the station downstream of Lyons Creek was predicted using RIVFLOC for various values of the cohesion parameter, β. The predictions were compared with measured size distributions for various values of β, and a value of 0.075 was chosen as an acceptable value. As can be seen from Figure 10, there is a reasonable agreement between the measured and predicted distributions for both smaller and larger flocs. However, for flocs in the intermediate size range (100 to 200 μm), the model over predicts the measured values somewhat. The percent deviation of the predicted floc size and the measured value ranged from $\pm$1.0% to $\pm$10%. The value of 0.075 for β is in the range of values that were measured in other similar studies that were carried out for sediments in Alberta (Droppo and Krishnappan [61]).

Table 4. Critical shear stresses for erosion and settling velocity of fine sediment from the Castle River watershed (Source: Stone et al. [62]).

Sediment Type	Consolidation Period (Days)	Settling Velocity mm/s	Critical Shear Stress for Erosion in Pa
	2	2.2	0.08
Burned	7	2.8	0.16
	14	3.0	0.18
	2	3.2	0.04
Unburned	7	3.3	0.10
	14	3.8	0.09

Table 5. Critical shear stresses for deposition of sediment (Krone's definition).

Sediment Type	Consolidation Period (Days)	Settling Velocity mm/s	Critical Shear Stress for Deposition in Pa
	2	2.2	0.40
Burned	7	2.8	0.64
	14	3.0	0.72
	2	3.2	0.20
Unburned	7	3.3	0.50
	14	3.8	0.45

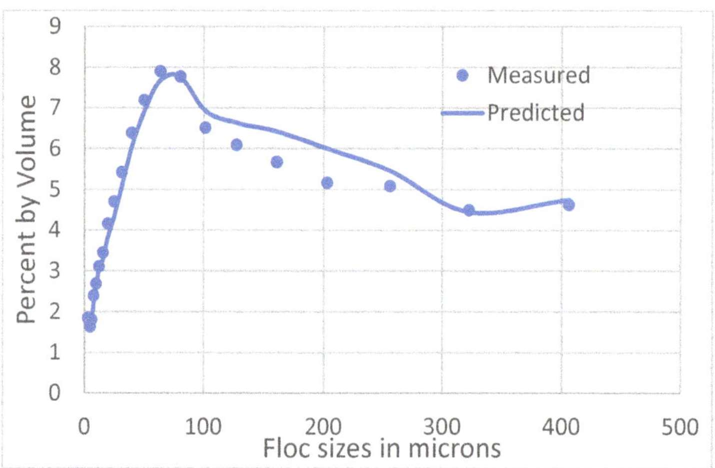

Figure 10. Comparison between measured size distribution and RIVFLOC prediction for the station downstream of the Lyons Creek confluence on the Crowsnest River.

2.2.9. Setting Up of TABS-MD for the Oldman Reservoir

TABS-MD (RMA2 and RMA4 with SMS user interface) was used to create a numerical grid based on reservoir bathymetry and to simulate both the flow field as well as fine sediment transport and dispersion within the Oldman Reservoir. To predict the flow pattern in the reservoir, an example stream flow condition of 25 m^3/s in the Crowsnest River, 75 m^3/s from the Oldman River and 100 m^3/s from the Castle River was used. These higher-than-average inflow rates were used to produce a sizable flow field that was expected to carry a significant sediment flux into the reservoir. The Manning's roughness coefficient for the reservoir was set as 0.025 (Hey [63]) and the turbulent eddy viscosity coefficient for the model was specified in terms of a Peclet Number with a value equal to 20 (Donnell et al. [58]). The flow patterns and velocity magnitude contours predicted by the model illustrate the complex flow patterns in the reservoir which are due to the irregular morphology and bathymetry. The prediction of sediment propagation over a 16-day period was evaluated using the model and is shown in Figure 11 to illustrate its capability. In this example, a sediment concentration of 1000 mg/L was used at the outlet of all three rivers. The model correctly predicts the flow rate through the dam thereby preserving flow continuity. Velocity in the reservoir ranged from 1 to 3 cm/s. Bed shear stresses predicted by the model range from 0.001 to 0.003 Pa and are two orders of magnitude lower than the critical shear stress for deposition of sediment shown in Table 5. Accordingly, the model predicts that most of the fine sediment entering the reservoir from all three rivers will be deposited within the reservoir and that resuspension of bottom sediment at normal operating water levels is unlikely.

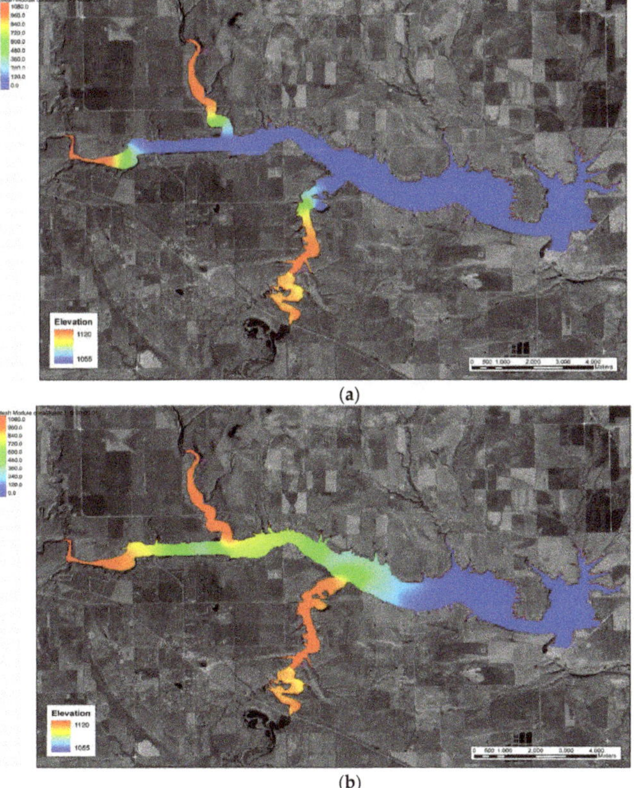

(a)

(b)

Figure 11. *Cont.*

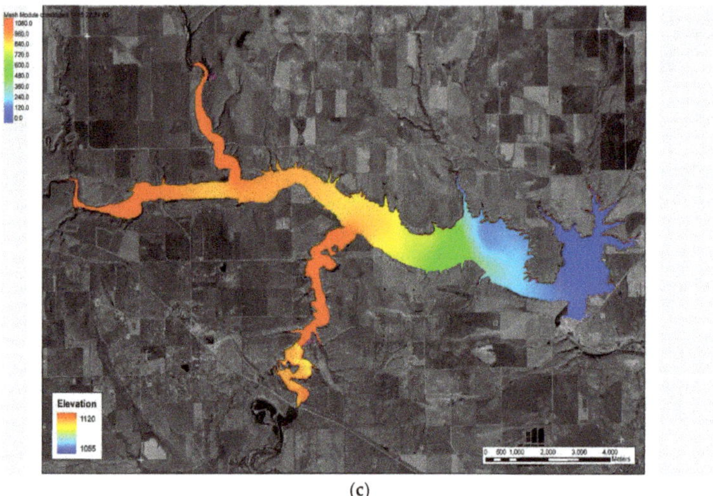

(c)

Figure 11. Simulation of sediment propagation through the Oldman Reservoir as predicted by TABS-MD ((**a**) Elapsed time = 2 days; (**b**) Elapsed time = 8 days; (**c**) Elapsed time = 16 days). The elevation legends shown as insets in these figures need to be ignored.

3. Results

The calibrated RIVFLOC model was run to estimate sediment flux for the Crowsnest River tributary to the Oldman reservoir (Figure 11). Predictions from RIVFLOC show that fine sediment in the Crowsnest River is transported directly to the Oldman reservoir if sediment entrapment in gravel beds is assumed to be zero (top horizontal line in Figure 12). However, several previous laboratory studies demonstrate that fine sediment deposition can occur in gravel beds because of the entrapment process (Einstein [68]; Packman et al. [69]; Rehg et al. [70]; Krishnappan and Engel [71]). The effect of coarse gravel on cohesive sediment entrapment was evaluated in an annular flume using fine sediment and gravel from the Elbow River in Alberta, which has similar land use characteristics (geology, vegetation, land use) to the Crowsnest River (Glasbergen et al. [72]). To evaluate the effect of entrapment on fine sediment transport dynamics in the Crowsnest River, two entrapment coefficients chosen from the literature (Krishnappan and Engel [71]; Glasbergen et al. [72]) were applied to RIVFLOC. The results show that the amount of sediment transported in the Crowsnest River decreases with distance downstream as the entrapment coefficient increases (Figure 12). Detailed field studies are required to quantify sediment entrapment dynamics in the Crowsnest River to refine model predictions. However, an entrapment coefficient (ratio between the entrapment flux and the settling flux) of 0.20 determined by Glasbergen et al. [72] provides a reasonable value and is used in this study to model fine sediment transport to the Oldman Reservoir. With this value of the entrapment coefficient, RIVFLOC predicts that 22% of the sediment entering the Crowsnest River will be entrapped within the river reach whereas the remaining 78% will be delivered downstream to the Oldman Reservoir.

RIVFLOC was also used to calculate the effective size distribution of solids at the downstream boundary of the Crowsnest River. Figure 13 compares this prediction with the size distribution of the sediment entering the river at the upstream boundary. From Figure 13, we can see that the effective size distribution at the downstream boundary increased with distance downstream due to flocculation. The predominant floc size at the upstream boundary was 70 μm but increased by a factor of 1.7 (~120 μm) at the downstream boundary. The effective sediment size distribution calculated from RIVFLOC for the downstream boundary, was used as the input for sediment size distribution for the reservoir model (RMA4).

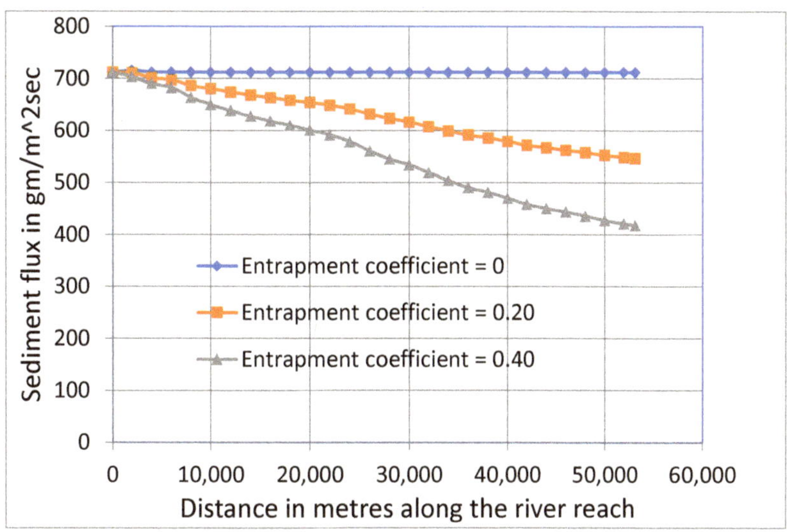

Figure 12. Fine sediment transport rate computed by RIVFLOC for the full Crowsnest River reach.

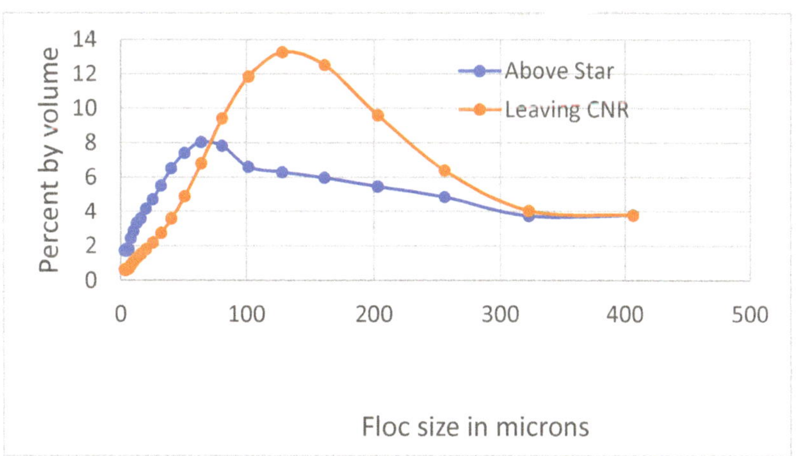

Figure 13. Comparison of effective particle size distributions of sediment entering and leaving the Crowsnest River reach.

RMA4 was used to determine the concentration of suspended sediment for individual size classes in the river inflow and at three transects (A, B, C) across the Oldman reservoir (Figure 1). Sediment deposition is simulated in RMA4 using the decay rate coefficient (k) which is equal to settling velocity divided by an average depth for any sediment size class. Simulated changes in sediment concentrations for various size fractions in different parts of the reservoir are presented in Figure 14. For this simulation, sediment inputs at the upstream boundary were held constant but different upstream boundary conditions in real time can also be used for scenario development or real time simulations. Previous research has demonstrated that the settling of flocs occurs at different rates for different size classes (Krishnappan [39]). RMA4 was used to simulate spatial variation in effective sediment size and concentration at the upstream, middle, and downstream sections of the reservoir (Figure 14). The data show that sediment in the size classes around 30 μm settles relatively quickly while both the finer (<30 μm) and larger sediment flocs (>100 μm) remain in suspension because of lower settling velocities. Given that bed shear stresses in the reservoir are two orders of magnitude lower than the critical shear stress for sediment deposition, the majority of suspended sediment entering the reservoir will deposit. The only size classes that are likely to be carried with the flow downstream of the Oldman dam are those that stay in the water column due to Brownian motion and the larger flocs that have very low settling velocities.

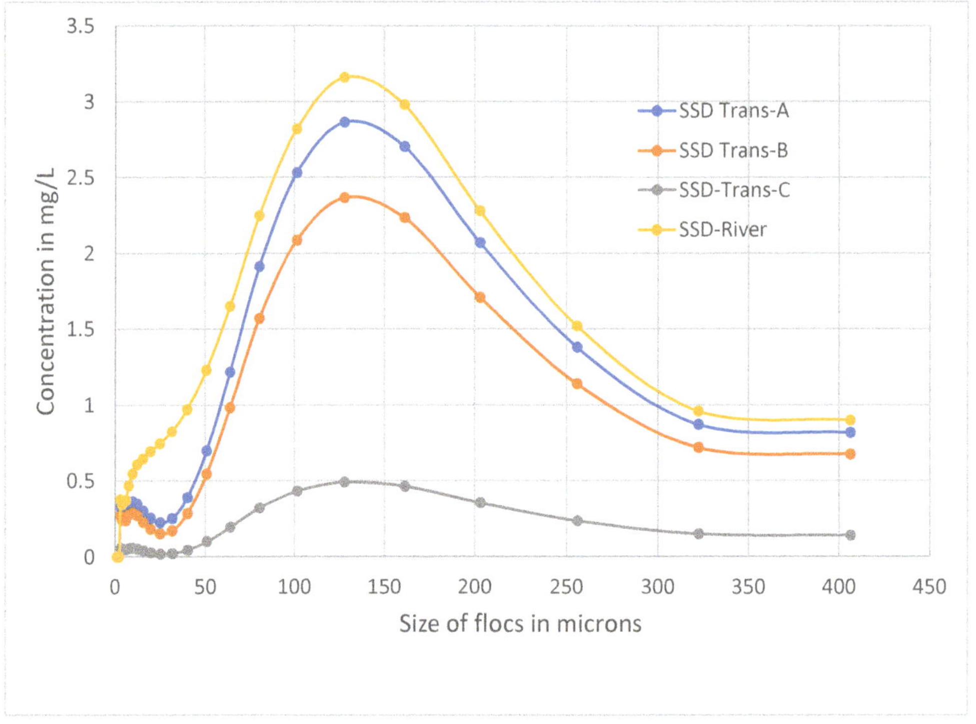

Figure 14. Effective particle size distributions of suspended sediment in river inflow and in the three transects across the Oldman Reservoir.

Simulating Sediment Flux to the Oldman Reservoir from Upstream Tributary Inflows to the Crowsnest River

While significant advancements in cohesive sediment transport models have been made, they are based primarily on the results of laboratory studies and are very seldom verified under field conditions particularly at large basin scales (Willis and Krishnappan [73]). One of the primary applications for the generic modelling framework described herein, is its utility for scenario development to explore specific risks of disturbance impacts (i.e., wildfire, extreme events, harvesting) on the propagation and fate of fine sediment in a reservoir from upstream sources. Here, we incorporate a five-year data set consisting of mean daily streamflow and suspended sediment concentration data as input in the modelling framework to simulate sediment transport dynamics in source water forested regions impacted by wildfire, simulating both propagation to, and through and, deposition within, a reservoir (Table 6). Simulations were made using an entrapment coefficient of 0.2 (Glasbergen et al. [72]) and results from the model are used to compare the amount and fate of suspended sediment from burned and post fire salvage logged watersheds (Lyons Creek) with that generated from predominantly unburned forested watersheds (Star Creek) and partially burned (South York Creek).

Table 6. Suspended sediment export (metric tonnes per year) from three Crowsnest River tributaries.

Tributaries	Sed. Type	2005	2006	2007	2008	2009	Total
Star Creek	unburned	3.26	26.66	12.98	14.38	5.91	63
South York	burned (56%)	2.57	21.83	80.73	71.59	27.33	204
Lyons Creek	burned	298.7	16.71	19.38	1406.87	46.05	1788
Total		304.7	65.2	113.09	1492.84	79.29	2055.11

The results of model simulations show differences in the depositional patterns for burned and unburned sediment deposited in four zones (Zone 1 is the region between the Crowsnest River entrance and the Transect A, Zone 2 is the region between Transect A and Transect B, Zone 3 is the region between the Transect B and Transect C and Zone 4 is the region between Transect C and the reservoir outlet) of the Oldman reservoir over the 5-year simulation period (Table 7).

Table 7. Sediment mass deposited in the Oldman Reservoir from burned and unburned tributary inflows.

Reservoir Zone	Unburned Sediment		Burned Sediment	
	Mass (t)	%	Mass (t)	%
1	39.3	33	385.8	26.0
2	16.1	13.5	210.7	14.2
3	51.4	43.1	712.2	48.0
4	12.4	10.4	175.1	11.8
Total	119.2	100	1483.8	100

The modelled sediment deposition patterns predicted by RMA4 demonstrate how differences in settling characteristics of the two sediment types influence dispersion and fate in the reservoir. The distinction between the behaviour of burned and unburned sediment is primarily due to the differences in corresponding settling velocities and this distinction is manifested in the reservoir because of the lower shear stresses that exist in this receptor. In the river channels, however, both burned and unburned sediment behave similarly because of the dominance of the bed shear stress over the small differences in the critical conditions between the two sediment types. Sediment generated from Lyons Creek (1788 tonnes) and partially burned South York Creek (204 × 0.56 = 114.2 tonnes) accounted for 93% of total sediment production (2055 tonnes) from the three monitored tributary inflows (Table 6). Out of the total amount of sediment produced, about 80% of the sediment is deposited in the reservoir (Table 7). To examine the impact of the 2003 Lost Creek wildfire on regional scale sediment production, Stone et al. [26] used a composite geochemical fingerprinting procedure to apportion the sediment efflux from three key spatial sediment sources: (1) unburned (reference), (2) burned and (3) burned sub-basins that were subsequently salvage logged. They reported that >80% of the downstream sediment contribution to the Oldman reservoir was produced from ~14% of the upstream landscape that was affected by the wildfire. The results of this source apportionment study agree favourably with modelled estimates of spatial sediment loading generated in the present study. Accordingly, the modelling framework can be accepted as providing planning and management level estimates of sediment delivery to the Oldman reservoir.

4. Discussion

The transport of cohesive sediment in aquatic systems is characterized by interactions among fine-grained primary sediment particles that cause flocs to form (Droppo [65]). Flocs have relatively low densities, large pore spaces and reactive surfaces that remove contaminants from the water column (Krishnappan [74]). The flocculation mechanism is dependent upon several factors including particle mineralogy, electrochemical nature of the flowing medium, biological factors such as bacteria and presence of other organic material and hydrodynamic properties of the flow field (Droppo et al. [75]). In a study of river and lake sediment, Droppo et al. [76] reported that only flocs < 100 μm (equivalent spherical diameter) settled within the Stokes' region (Re < 0.2). The densities of these flocculated materials ranged from 1 to 1.4 g cm^{-3} but the majority of flocs had densities of less than 1.1 g cm^{-3}. Floc porosity increases with floc size and low floc densities are caused by the entrapment of water in the pore spaces of flocs (Droppo et al. [75]). Accordingly, flocculation is an important mechanism for particle removal in aquatic environments such as streams and reservoirs because it alters the hydrodynamic characteristics of solids by changing their density, porosity, settling velocity and surface area (Willis and Krishnappan [73]). In the present study, flocculation altered both the effective particle size distribution and dispersion of suspended sediment along a 50 km reach of the Crowsnest River (Figure 13) and in the Oldman reservoir (Figure 14). It should be underscored that RMA4 does not presently take flocculation into account and hence the flocculation of the sediment that can occur within the reservoir was neglected. Efforts are underway to incorporate the flocculation module of RIVFLOC into RMA4.

Since elevated sediment loss is a pervasive problem associated with several types of land disturbances, the environmental significance and ecological importance of fine sediment is increasingly being recognized as a critical component of watershed management. For example, the Water Framework Directive of the European Union recognises the need for fine sediment management albeit in the context of ongoing scientific debate on the most appropriate compliance targets (European Parliament [76]; Collins and Anthony [77]). From a traditional hydraulic engineering perspective, fine sediment transport was considered to have limited importance for river morphology, channel sedimentation or reservoir management (Walling and Collins [30]). However, viewed from more transdisciplinary and especially ecological and contaminant transport perspectives, improved knowledge of the nature, mobility and transport dynamics of fine sediment and application of this information for model development and use is critical to develop and refine robust management tools for the protection of water supplies, aquatic ecology, and riparian systems under current or future climate change scenarios (Ice et al. [78]).

Improved description of the nature, mobility and transport dynamics of fine sediment is especially important in gravel bed rivers such as those in forested headwater regions with historically low sediment yields because they are often the most sensitive to minor changes in fine sediment inputs (Walling and Collins [30]; Watt et al. [79]). The significance of sediment particle size for the redistribution and fate of contaminant transport in aquatic systems has been widely reported (e.g., Ongley et al. [28]; Walling and Woodward [80]). It is a key factor controlling the form, mobility, transport, and dispersion of sediment-associated contaminants (Horowitz and Elrick [27]; Stone and Mudroch [81]; Stone and English [82]). Several studies on the effects of wildfire on nutrient export in streams show that postfire phosphorus (P) export can increase from 0.3 to more than five times greater than at pre-fire conditions (Blake et al. [83]; Silins et al. [20]). In some locations, post-fire recovery in P export can occur within two years while elsewhere a more prolonged recovery has been observed (Silins et al. [20]). For example, in an investigation of P speciation and sorption behavior of suspended sediment affected by the Lost Creek wildfire in the Crowsnest River, Emelko et al. [30] reported that sediments from the burned tributary inputs contained higher levels of bioavailable particulate P (NAIP) which were observed downstream at larger river basin scales. They also found that the potential of burned sediment to release P to the water column was significantly higher downstream of wildfire-impacted areas compared to sediment from upstream reference (unburned) river reaches. Notably, approximately 80% of the fine post-fire sediment deposited in the downstream Oldman Reservoir originated from only ~14% of the total watershed area (Stone et al. [26]). Collectively, these previous reports emphasize the critical importance of understanding sediment transport from burned landscapes to the Crowsnest River: the elevated post-fire concentrations of bioavailable NAIP are preferentially bound to, and carried by, fine sediment, and have significant implications for the quality and treatability of water stored in the Oldman reservoir.

The mobilization and transport of sediment-associated bioavailable NAIP represents two particular water quality and treatability problems—it is critical to recognize that may manifest over different time scales. First, fine sediment entering the reservoir may remain in suspension for days; however, differences in soluble reactive phosphorus (SRP) concentrations in the river and reservoir may cause the desorption of P from the sediment into the reservoir water column, especially if the fine sediment is rich in P as has been observed after severe wildfire (Emelko et al. [21]; Watt et al. [79]). This process can occur over a period of hours (Froelich [84]). Second, over longer time periods, P-enriched sediment deposited in the reservoir may serve as a source of internal P loading to the reservoir water column because SRP can be released from anoxic sediments at the sediment-water interface (Nürnberg [85]). Accordingly, both suspended and deposited sediment can be sources of bioavailable P that can promote algal proliferation, which can lead to the associated production cyanotoxins of health concern or unpleasant taste and odour compounds (Emelko et al. [5,21]). Accordingly, the flushing frequency of the reservoir may have to be increased for risk management in response to continued loading of P-enriched fine sediment to the reservoir from wildfire-impacted upstream landscapes. Notably, this practice may likely have further implications for downstream water quality and aquatic ecology due to the more frequent release of nutrient-rich fine sediment, which may include benthic invertebrate community structure, invertebrate density, biomass, species diversity, and shifts in species composition (Silins et al. [20]; Martens et al. [86]).

5. Conclusions

This study demonstrates key features of a new integrated modelling framework (MOBED, RIVFLOC, RMA2, RMA4) to quantify cohesive sediment fluxes to reservoirs and inform post-fire reservoir management. While the framework was applied herein to demonstrate the prediction of fine sediment transport and fate in wildfire impacted rivers, it can be applied to inform watershed risk management and drinking water source protection by describing downstream sediment dynamics resulting from a broad range of land disturbance scenarios. The principal conclusions from this work are:

1. A new integrated modelling framework to quantify sediment fluxes to reservoirs was developed and validated. It is the first such platform for describing fine sediment transport that includes explicit description of fine sediment deposition/erosion processes as a function of bed shear stress *and* the flocculation process.
2. Bed shear stresses that prevail even at low flow conditions in the study reaches are considerably higher than the critical shear stress for deposition of fine sediment generated in the watershed. This indicates that most of the fine sediment entering the Crowsnest River from tributary inflows will be readily transported through the channel network to the downstream reservoir.
3. The process of flocculation changes the particle size distribution of suspended sediment in the water column of the Crowsnest River and influences the dispersion pattern of particles in the Oldman Reservoir because flocculation impacts the settling velocity, porosity, and density of aggregated particles (i.e., flocs). Thus, this process is an essential component of any fine (i.e., cohesive) sediment transport model, as demonstrated and validated herein. Exclusion of the flocculation process can result in underestimation of fine sediment and associated contaminant transport.

4. Deposition due to entrapment in the gravel bed study river is a possibility and this process needs to be examined further to support new process-based model parameterization.
5. Deposition patterns of sediment from wildfire-impacted landscapes were different than those from unburned landscapes because of differences in settling behaviour. These differences may lead to zones of relatively increased internal loading of phosphorus to reservoir water columns, thereby increasing the potential for algae proliferation.

Author Contributions: Conceptualization, B.G.K., M.S., A.L.C. and U.S.; methodology, B.G.K., M.S., U.S. and A.L.C.; software, validation, B.G.K. and C.H.S.W.; formal analysis, B.G.K. and M.S.; investigation, resources, C.H.S.W.; data curation, B.G.K., S.A.S. and U.S.; writing—original draft preparation, M.S. and B.G.K.; writing—review and editing, M.S., B.G.K., M.B.E., A.L.C. and U.S.; visualization, B.G.K.; supervision, M.S. and U.S.; project administration, M.S. and B.G.K.; funding acquisition, M.S., U.S., M.B.E. and A.L.C. All authors have read and agreed to the published version of the manuscript.

Funding: Field work and lab analyses were funded by NSERC Discovery Grant 481 RGPIN-2020-06963 awarded to M. Stone; Alberta Innovates Energy and Environment Solutions Grant AI-EES:2096 awarded to U. Silins, M.B. Emelko and M. Stone; and Alberta Innovates BIO Grant AI-BIO: Bio-13-009 awarded to U. Silins, M.B. Emelko and M. Stone. MBE's contribution was also enabled in part, thanks to funding from the Canada Research Chairs (CRC) Program. The contribution of ALC to this work was funded in part by the UK Biotechnology and Biological Sciences Research Council (UKRI-BBSRC) via grant BBS/E/C/000I0330. Contributions by US, CHSW, and SAS were supported by grants from ESRD/AAF (13GRFM15,15GRFFM11), AI (1865, 2343), and NSERC (216984).

Data Availability Statement: Data supporting reported results are with the authors can be made available upon request.

Acknowledgments: We gratefully acknowledge field assistance by Amanda Martens, Ashley Peter-Rennich, Samantha Karpyshin, Hans Biberhofer, Amelia Corrigan, and Kalli Herlein.

Conflicts of Interest: The authors declare no conflict of interest.

References

1. Caldwell, P.; Muldoon, C.; Ford-Miniat, C.; Cohen, E.; Krieger, S.; Sun, G.; McNulty, S.; Bolstad, P.V. *Quantifying the Role of National Forest System Lands in Providing Surface Drinking Water Supply for the Southern United States*; Gen. Technical Report SRS-197; U.S. Department of Agriculture Forest Service, Southern Research Station: Asheville, NC, USA, 2014; 135p.
2. Kastridis, A.; Kamperidou, V. Influence of land use changes on alluviation of Volvi Lake wetland (North Greece). *Soil Water Res.* **2015**, *10*, 121–129. [CrossRef]
3. Vacca, A.; Loddo, S.; Ollesch, G.; Puddu, R.; Serra, G.; Tomasi, D.; Aru, A. Measurement of runoff and soil erosion in three areas under different land use in Sardinia (Italy). *Catena* **2000**, *40*, 69–92. [CrossRef]
4. Vörösmarty, C.J. Global Water Resources: Vulnerability from Climate Change and Population Growth. *Science* **2000**, *80*, 284–288. [CrossRef]
5. Emelko, M.B.; Silins, U.; Bladon, K.D.; Stone, M. Implications of land disturbance on drinking water treatability in a changing climate: Demonstrating the need for "source water supply and protection" strategies. *Water Res.* **2011**, *45*, 461–472. [CrossRef]
6. IPCC. Climate Change 2014: Impacts, Adaptation, and Vulnerability. In *Part A: Global and Sectorial Aspects*; Contribution of the Working Group II to the Fifth Assessment Report of the IPCC; IPCC: Geneva, Switzerland, 2014; 1132p.
7. Vörösmarty, C.J.; McIntyre, P.B.; Gessner, M.O.; Dudgeon, D.; Prusevich, A.; Green, P.A.; Glidden, S.; Bunn, S.E.; Sullivan, C.A.; Liermann, C.R.; et al. Global threats to human water security and river biodiversity. *Nature* **2010**, *467*, 555–561. [CrossRef]
8. Emelko, M.B.; Sham, C.H. *Wildfire Impacts on Water Supplies and the Potential for Mitigation: Workshop Report*; Canadian Water Network and Water Research Foundation: Waterloo, ON, Canada, 2014.
9. Vose, J.M.; Sun, G.; Ford, C.R.; Bredemeier, M.; Otsuki, K.; Wei, X.; Zhang, Z.; Zhang, L. Forest eco-hydrological research in the 21st century: What are the critical needs? *Ecohydrology* **2011**, *4*, 146–158. [CrossRef]
10. Khan, S.J.; Deere, D.; Leusch, F.D.L.; Humpage, A.; Jenkins, M.; David Cunliffe, D. Extreme weather events: Should drinking water quality management systems adapt to changing risk profiles. *Water Res.* **2015**, *85*, 124–136. [CrossRef] [PubMed]
11. Westerling, A.L.; Hidalgo, H.G.; Cayan, D.R.; Swetnam, T.W. Warming and earlier spring increase western U.S. forest wildfire activity. *Science* **2006**, *313*, 940–943. [CrossRef] [PubMed]
12. Flannigan, M.D.; Wotton, M.; Marshall, G.; de Groot, W.J.; Johnston, J.; Jurko, N.; Cantin, A.S. Fuel moisture sensitivity to temperature and precipitation: Climate change implications. *Clim. Chang.* **2016**, *134*, 59–71. [CrossRef]

13. Benda, L.; Miller, D.; Bigelow, P.; Andras, K. Effects of post-wildfire erosion on channel environments, Boise River, Idaho. *For. Ecol. Manag.* **2003**, *178*, 105–119. [CrossRef]
14. Kinoshita, A.M.; Chin, A.; Simon, G.L.; Briles, C.; Hogue, T.S.; O-Dowd, A.P.; Gerlak, A.K.; Albornoz, A.U. Wildfire, water and society: Toward integrative research in the 'anthropocene'. *Anthropocene* **2016**, *16*, 16–27. [CrossRef]
15. Bisson, P.A.; Rieman, B.E.; Luce, C.; Hessburg, P.F.; Lee, D.C.; Kershner, J.L.; Reeves, G.H.; Gresswell, R.E. Fire and aquatic ecosystems of the western USA: Current knowledge and key questions. *For. Ecol. Manag.* **2003**, *178*, 213–229. [CrossRef]
16. Silins, U.; Bladon, K.; Stone, M.; Emelko, M.; Boon, S.; Williams, C.; Wagner, M.; Howery, J. *Impacts of Natural Disturbance by Wildfire on Hydrology, Water Quality and Aquatic Ecology of Rocky Mountain Watersheds—Phase I (2004–2008)*; Report on Southern Rockies Watershed Project; Alberta Sustainable Resource Development: Edmonton, AB, Canada, 2009; 90p.
17. Lucas-Borja, M.; Zema, D.A.; Plaza-Álvarez, P.; Zupanc, V.; Baartman, J.; Sagra, J.; González-Romero, J.; Moya, D.; de las Heras, J. Effects of different land uses (abandoned farmland, intensive agriculture and forest) on soil hydrological properties in southern Spain. *Water* **2019**, *11*, 503. [CrossRef]
18. Plaza-Álvarez, P.A.; Lucas-Borja, M.E.; Sagra, J.; Zema, D.A.; González-Romero, J.; Moya, D.; De las Heras, J. Changes in soil hydraulic conductivity after prescribed fires in Mediterranean pine forests. *J. Environ. Manag.* **2019**, *232*, 1021–1027. [CrossRef] [PubMed]
19. Emmerton, C.A.; Cooke, C.A.; Hustins, S.; Silins, U.; Emelko, M.B.; Lewis, T.; Kruk, M.K.; Taube, N.; Zhu, D.; Jackson, B.; et al. Severe western Canadian wildfire affects water quality even at large basin scales. *Water Res.* **2020**, *183*, 116071. [CrossRef] [PubMed]
20. Silins, U.; Bladon, K.D.; Kelly, E.N.; Esch, E.; Spence, J.R.; Stone, M.; Emelko, M.B.; Boon, S.; Wagner, M.J.; Williams, C.H.S.; et al. Five-year legacy of wildfire and salvage logging impacts on nutrient runoff and aquatic plant, invertebrate, and fish productivity. *Ecohydrology* **2014**, *7*, 1508–1523. [CrossRef]
21. Emelko, M.B.; Stone, M.; Silins, U.; Allin, D.; Collins, A.L.; Williams, C.H.; Martens, A.M.; Bladon, K.D. Sediment-phosphorus dynamics can shift aquatic ecology and cause downstream legacy effects after wildfire in large river systems. *Glob. Chang. Biol.* **2016**, *22*, 1168–1184. [CrossRef] [PubMed]
22. Tobergte, D.R.; Curtis, S. The effects of wildfire on stream ecosystems in the western United States: Magnitude, persistence, and factors affecting recovery. In Proceedings of the World Environmental and Water Resources Congress 2016, West Palm Beach, FL, USA, 22–26 May 2016. [CrossRef]
23. López-Vicente, J.; González-Romero, M.E. Lucas-Borja 2020. Forest fire effects on sediment connectivity in headwater sub-catchments: Evaluation of indices performance. *Sci. Total Environ.* **2020**, *732*, 139205. [CrossRef]
24. Malmon, D.V.; Reneau, S.L.; Katzman, D.; Lavine, A.; Lyman, J. Suspended sediment transport in an ephemeral stream following wildfire. *J. Geophys. Res.* **2007**, *112*, F02006. [CrossRef]
25. Silins, U.; Stone, M.; Emelko, M.B.; Bladon, K.D. Sediment production following severe wildfire and post-fire salvage logging in the Rocky Mountain headwaters of the Oldman River Basin, Alberta. *Catena* **2009**, *79*, 189–197. [CrossRef]
26. Stone, M.; Collins, A.; Silins, U.; Emelko, M.B.; Zhang, Y. The Use of Composite Fingerprints to Quantify Sediment Sources in a Wildfire Impacted Forested Landscape, Alberta, Canada. *Sci. Total Environ.* **2014**, *473–474*, 642–650. [CrossRef]
27. Horowitz, A.J.; Elrick, K. The relation of stream sediment surface area, grain size and composition to trace element chemistry. *Appl. Geochem.* **1987**, *2*, 437–451. [CrossRef]
28. Ongley, E.; Krishnappan, B.G.; Droppo, I.; Rao, S.S.; Maguire, R.J. Cohesive sediment transport: Emerging issues for toxic chemical management. *Hydrobiologia* **1992**, *235–236*, 177–187. [CrossRef]
29. Chapman, P.M.; Wang, F.; Caeiro, S. Assessing and managing sediment contamination in transitional waters. *Environ. Int.* **2013**, *55C*, 71–91. [CrossRef] [PubMed]
30. Walling, D.E.; Collins, A. Fine sediment transport and management. In *River Science: Research and Management for the 21st Century*, 1st ed.; Gilvear, D.J., Greenwood, M.T., Thoms, M.C., Wood, P.J., Eds.; John Wiley & Sons, Ltd.: Hoboken, NJ, USA, 2016.
31. Collins, A.L.; Naden, P.S.; Sear, D.A.; Jones, J.I.; Foster, I.D.L.; Morrow, K. Sediment targets for informing river catchment management: International experience and prospects. *Hydrol. Process.* **2011**, *25*, 2112–2129. [CrossRef]
32. Wood, P.; Armitage, D. Biological Effects of Fine Sediment in the Lotic Environment. *Environ. Manag.* **1997**, *21*, 203–217. [CrossRef] [PubMed]
33. Partheniades, E. Results of recent investigations on erosion and deposition of cohesive sediments. *Sedimentation* **1972**, 20–21.
34. Krone, R.B. Flume studies of the transport of sediment in estuarial shoaling processes. In *Final Report, Hydraulic Engineering Laboratory and sanitary Engineering Research Laboratory*; University of California: Berkeley, CA, USA, 1962.
35. Mehta, A.J. (Ed.) Characterization cohesive sediment properties and transport processes in estuaries. In *Estuarine Sediment Dynamics*; Lecture Notes on Coastal and Estuarine Studies; Springer: New York, NY, USA, 1986; Volume 14, pp. 290–325.
36. Lick, W. Entrainment, deposition and transport of fine-grained sediments in lakes. *Hydrobiologia* **1982**, *91*, 31–40. [CrossRef]
37. Lau, Y.L.; Krishnappan, B.G. Does re-entrainment occur during cohesive sediment settling? *J. Hydraul. Eng. ASCE* **1994**, *120*, 236–244. [CrossRef]
38. Krishnappan, B.G. *A New Algorithm for Fine Sediment Transport*; Contribution No. 96–160; National Water Research Institute, Environment Canada: Burlington, ON, Canada, 1996.
39. Krishnappan, B.G. Recent advances in basic and applied research on cohesive sediment transport in aquatic systems. *Can. J. Civil Eng.* **2007**, *34*, 731–743. [CrossRef]

40. Partheniades, E.; Kennedy, J.F. Depositional Behaviour of Fine Sediment in a Turbulent Fluid Motion. In Proceedings of the 10th Conference on Coastal Engineering, Tokyo, Japan, 29 January 1966; Volume 2, pp. 708–724.

41. Mehta, A.J.; Partheniades, E. An investigation of the depositional properties of flocculated fine sediments. *J. Hydraul. Res.* **1975**, *13*, 361–381. [CrossRef]

42. Rodrigues, E.L.; Jacobi, C.M.; Figueira, J.E.C. Wildfires and their impact on the water supply of a large neotropical metropolis: A simulation approach. *Sci. Total Environ.* **2019**, *651*, 1261–1271. [CrossRef] [PubMed]

43. Moody, J.A.; Shakesby, R.A.; Robichaud, P.R.; Cannon, S.H.; Martin, D.A. Current research issues related to post-wildfire runoff and erosion processes. *Earth-Sci. Rev.* **2013**, *122*, 10–37. [CrossRef]

44. Lancaster, S.T.; Hayes, S.K.; Grant, G.E. Modeling sediment and wood storage and dynamics in small mountainous watersheds. In *Geomorphic Processes and Riverine Habitat*; American Geophysical Union: Washington, DC, USA, 2001; pp. 85–102.

45. Istanbulluoglu, E.; Tarboton, D.G.; Pack, R.T.; Luce, C.H. Modelling of the interactions between forest vegetation, disturbances, and sediment yields. *J. Geophys. Res.* **2004**, *109*, F01009. [CrossRef]

46. Langhans, C.; Smith, H.G.; Chong, D.M.O.; Nyman, P.; Lane, P.N.J.; Sheridan, G.J. A model for assessing water quality risk in catchments prone to wildfire. *J. Hydrol.* **2016**, *534*, 407–426. [CrossRef]

47. Santos, R.M.B.; Sanches Fernandes, L.F.; Pereira, M.G.; Cortes, R.M.V.; Pacheco, F.A.L. Water resources planning for a river basin with recurrent wildfires. *Sci. Total Environ.* **2015**, *526*, 1–13. [CrossRef] [PubMed]

48. Danish Hydraulic Institute. MIKE Hydro Basin Water Quality Model. ECOLab Template Documentation, Scientific Documentation. Water and Environment, Danish Hydraulic Institute, Hørsholm. 2018. Available online: https://manuals.mikepoweredbydhi.help/2019/Water_Resources/MIKEHYDRO_BasinWQ_Scientific.pdf (accessed on 15 August 2021).

49. Danish Hydraulic Institute. MIKE 1D Sediment Transport, Scientific Documentation. Water and Environment, Danish Hydraulic Institute, Hørsholm. 2020. Available online: https://manuals.mikepoweredbydhi.help/latest/Water_Resources/MIKE1D_ST_ScientificManual.pdf (accessed on 15 August 2021).

50. Anand, J.; Gosain, A.K.; Khosa, R. Impacts of climate and land use change on hydrodynamics and sediment transport regime of the Ganga River Basin. *Reg. Environ. Chang.* **2021**, *21*, 1–5. [CrossRef]

51. Walling, D.E.; Wilkinson, S.N.; Horowitz, A.J. Catchment Erosion, Sediment Delivery, and Sediment Quality. In *Treatise on Water Science*; Wilderer, P., Ed.; Academic Press: Oxford, UK, 2011; Volume 2, pp. 305–338.

52. Daniel, E.; Camp, J.V.; LeBoeuf, E.J.; Penrod, J.R.; Dobbins, J.P.; Abkowitz, M.D. Watershed Modeling and its Applications: A State-of-the-Art Review. *Open Hydrol. J.* **2011**, *5*, 26–50. [CrossRef]

53. Summer, W.; Walling, D.E. *Modelling Erosion, Sediment Transport and Sediment Yield*; SC-2002/WS/48; UNESCO: Paris, France, 2002.

54. Krishnappan, B.G. *User Manual: Unsteady, Non-Uniform Mobile Boundary Flow Model-MOBED*; Hydraulics Division, National Water Research Institute: Burlington, ON, Canada, 1981; 197p.

55. Krishnappan, B.G. *MOBED User Manual, Update I*; Hydraulics Division, National Water Research Institute: Burlington, ON, Canada, 1983; 95p.

56. Krishnappan, B.G. *MOBED User Manual, Update II*; Hydraulics Division, National Water Research Institute: Burlington, ON, Canada, 1986; 95p.

57. Krishnappan, B.G. Modelling of cohesive sediment transport. In Proceedings of the International Symposium on the Transport of Suspended Sediments and Its Mathematical Modelling, Florence, Italy, 2–5 September 1991; pp. 433–448.

58. Donnell, B.P.; Letter, J.V.; McNally, W.H. *User Guide for RMA2 Version 4.5*; U.S. Army, Engineer Research and Development Centre, Waterways Experimental Station, Coastal and Hydraulics Laboratory: Valhalla, NY, USA, 2008; 279p.

59. Letter, J.V.; Donnell, B.P.; Brown, G.L. *User Guide for RMA4 Version 4.5*; U.S. Army, Engineer Research and Development Centre, Waterways Experimental Station, Coastal and Hydraulics Laboratory: Valhalla, NY, USA, 2008; 279p.

60. Krishnappan, B.G.; Marsalek, J. Transport characteristics of fine sediments from an on-stream stormwater management pond. *Urban Water* **2002**, *4*, 3–11. [CrossRef]

61. Droppo, I.G.; Krishnappan, B.G. Modeling of hydrophobic cohesive sediment transport in the Ells River Alberta Canada. *J. Soils Sediments* **2016**, *16*, 2753–2765. [CrossRef]

62. Stone, M.; Emelko, M.B.; Droppo, I.G.; Silins, U. Bio-stabilization and erodibility of cohesive sediment deposits in wildfire affected streams. *Water Res.* **2010**, *45*, 521–534. [CrossRef]

63. Hey, R.D.; Bathurst, J.C.; Thorne, C.R. *Gravel Bed Rivers*; John Wiley and Sons: New York, NY, USA, 1982.

64. Lau, Y.L.; Krishnappan, B.G. *Measurement of Size Distribution of Settling Flocs*; NWRI Contribution No. 97-223; Environment Canada: Burlington, ON, Canada, 1997; 21p.

65. Droppo, I.G. Rethinking what constitutes suspended sediment. *Hydrol. Process.* **2001**, *15*, 1551–1564. [CrossRef]

66. Krishnappan, B.G.; Stephens, R. *Critical Shear Stresses for Erosion and Deposition of Suspended Sediment from the Athabasca River*; Report No. 85; Northern River Basins Study: Edmonton, AB, Canada, 1995.

67. Krishnappan, B.G.; Stephens, R.; Moore, B.; Kraft, J. *Transport Characteristics of Fine Sediments in the Fraser River System*; NWRI Contribution No. 97-220; National Water Research Institute, Environment Canada: Burlington, ON, Canada, 1997; 21p.

68. Einstein, H.A. *The Bed Load Function for Sediment Transportation in Open Channel Flows*; Technical Bulletin No. 1026; U.S. Department of Agriculture, Soil Conservation Service: Washington, DC, USA, 1950.

69. Packman, A.I.; Brooks, N.H.; Morgan, J.J. Kaolinite exchange between a stream and streambed: Laboratory experiments and validation of a colloid transport model. *Water Resour. Res.* **2000**, *36*, 2351–2361. [CrossRef]

70. Rehg, K.; Packham, A.I.; Ren, J. Effects of suspended sediment characteristics and bed sediment transport on streambed clogging. *Geology* **2005**, *19*, 413–427. [CrossRef]
71. Krishnappan, B.G.; Engel, P. Entrapment of fines in coarse sediment beds. In *River Flow 2006*; Alves, E.C.T.L., Ferreira, R.M.L., Leal, J.G.A.B., Cardoso, A.H., Eds.; Tailor and Francis Group: London, UK, 2006; pp. 817–824.
72. Glasbergen, K.; Stone, M.; Krishnappan, B.; Dixon, J.; Silins, U. The effect of coarse gravel on cohesive sediment entrapment in an annular flume. In *Sediment Dynamics from the Summit to the Sea, 2014*; (IAHS Publ. 367); IAHS: Wallingford, UK, 2014.
73. Willis, D.; Krishnappan, B.G. Numerical modeling of cohesive sediment transport in rivers. *Can. J. Civ. Eng.* **2004**, *31*, 749–758. [CrossRef]
74. Krishnappan, B.G. Erosion behaviour of fine sediment deposits. *Can. J. Civ. Eng.* **2004**, *31*, 759–766. [CrossRef]
75. Droppo, I.G.; Lepard, G.G.; Liss, S.N.; Milligan, T.G. *Flocculation in Natural and Engineered Environmental Systems*; CRC Press: Boca Raton, FL, USA, 2004; 438p.
76. European Parliament. *Establishing a Framework for Community Action in the Field of Water Policy*; Directive, EC/2000/60; European Parliament: Brussels, Belgium, 2000.
77. Collins, A.L.; Anthony, S.G. Assessing the likelihood of catchments across England and Wales meeting 'good ecological status' due to sediment contributions from agricultural sources. *Environ. Sci. Policy* **2008**, *11*, 163–170. [CrossRef]
78. Ice, G.; Neary, D.; Adams, P. Effects of Wildfire on Soils and Watershed Processes. *J. For.* **2004**, *102*, 16–20.
79. Watt, C.; Emelko, M.B.; Silins, U.; Collins, A.L.; Stone, M. Anthropogenic and climate-exacerbated landscape disturbances converge to alter phosphorus bioavailability in an oligotrophic river under pressure. *ChemRxiv* **2021**.
80. Walling, D.E.; Woodward, J.C. Use of a field-based water elutriation system for monitoring the in-situ particle size characteristics of fluvial suspended sediment. *Water Res.* **1993**, *27*, 1413–1421. [CrossRef]
81. Stone, M.; Mudroch, A. The effect of particle size, chemistry and mineralogy of river sediments on phosphate adsorption. *Environ. Technol. Lett.* **1989**, *10*, 501–510. [CrossRef]
82. Stone, M.; English, M.C. Geochemical composition, phosphorus speciation and mass transport characteristics of fine-grained sediment in two Lake Erie tributaries. *Hydrobiologia* **1993**, *253*, 17–29. [CrossRef]
83. Blake, W.H.; Theocharopoulos, S.P.; Skoulikidis, N.; Clark, P.; Tountas, P.; Hartley, R.; Amaxidis, Y. Wildfire impacts on hillslope sediment and phosphorus yields. *J. Soils Sediments* **2010**, *10*, 671–682. [CrossRef]
84. Froelich, P.N. Kinetic control of dissolved phosphate in natural rivers and estuaries: A primer on the phosphate buffer mechanism. *Limnol. Oceanogr.* **1988**, *33*, 649–668. [CrossRef]
85. Nürnberg, G.K. Assessing internal phosphorus load—Problems to be solved. *Lake Reserv. Manag.* **2009**, *25*, 419–432. [CrossRef]
86. Martens, A.M.; Silins, U.; Proctor, H.C.; Williams, C.H.S.; Wagner, M.J.; Emelko, M.B.; Stone, M. Long term impact of severe wildfire on macroinvertebrate assemblage structure in Alberta's Rocky Mountains. *Int. J. Wildland Fire* **2019**, *28*, 738–749. [CrossRef]

Article

3D Numerical Simulation of Gravity-Driven Motion of Fine-Grained Sediment Deposits in Large Reservoirs

Dongdong Jia [1], Jianyin Zhou [2,*], Xuejun Shao [3] and Xingnong Zhang [1]

[1] State Key Laboratory of Hydrology-Water Resources and Hydraulic Engineering, Nanjing Hydraulic Research Institute, Nanjing 210029, China; ddjia@nhri.cn (D.J.); xnzhang@nhri.cn (X.Z.)
[2] Changjiang River Scientific Research Institute, Wuhan 430010, China
[3] State Key Laboratory of Hydroscience and Engineering, Tsinghua University, Beijing 100084, China; shaoxj@mail.tsinghua.edu.cn
* Correspondence: zhoujianyin@mail.crsri.cn

Abstract: Deposits in dam areas of large reservoirs, which are commonly composed of fine-grained sediment, are important for reservoir operation. Since the impoundment of the Three Gorges Reservoir (TGR), the sedimentation pattern in the dam area has been unexpected. An integrated dynamic model for fine-grained sediment, which consists of both sediment transport with water flow and gravity-driven fluid mud at the bottom, was proposed. The incipient motion driven by gravity in the form of fluid mud was determined by the critical slope. Shallow flow equations were simplified to simulate the gravity-driven mass transport. The gravity-driven flow model was combined with a 3D Reynolds-averaged water flow and sediment transport model. Solution routines were developed for both models, which were then used to simulate the integral movement of the fine-grained sediment. The simulated sedimentation pattern agreed well with observations in the dam area of the TGR. Most of the deposits were found at the bottom of the main channel, whereas only a few deposits remained on the bank slopes. Due to the gravity-driven flow of fluid mud, the deposits that gathered in the deep channel formed a nearly horizontal surface. By considering the gravity-driven flow, the averaged error of deposition thickness along the thalweg decreased from −13.9 to 2.2 m. This study improved our understanding of the mechanisms of fine-grained sediment transport in large reservoirs and can be used to optimize dam operations.

Keywords: dam area of reservoirs; fine-grained sediment; fluid mud; gravity-driven flow; reservoir sedimentation

Citation: Jia, D.; Zhou, J.; Shao, X.; Zhang, X. 3D Numerical Simulation of Gravity-Driven Motion of Fine-Grained Sediment Deposits in Large Reservoirs. *Water* **2021**, *13*, 1868. https://doi.org/10.3390/w13131868

Academic Editor: Bommanna Krishnappan

Received: 9 May 2021
Accepted: 29 June 2021
Published: 4 July 2021

Publisher's Note: MDPI stays neutral with regard to jurisdictional claims in published maps and institutional affiliations.

1. Introduction

The impoundment of large reservoirs results in numerous changes, including decreases in water flow velocity, sediment deposition, and sorting [1]. In the reservoir, coarser sediments tend to be deposited in the upstream regions of the reservoir, advancing downstream steadily but slowly in the form of a delta, whereas fine sediments reach the reservoir in the form of suspended load, by means of homogeneous or stratified dense flow [2].

Sedimentation is an important issue that receives a significant amount of attention in reservoir construction and operations [3]. The excess of sedimentation in a reservoir leads to sediment entrainment in waterway systems and hydraulic schemes [4]. During the demonstration phase and early operational periods of the Three Gorges Reservoir (TGR), several hydraulic research institutes studied the problem of reservoir sedimentation [5,6]. After the impoundment, the sediment influx conditions of the TGR changed significantly, and the actual reservoir sedimentation conditions were different from the predictions made during the demonstration phase [3,7]. With the increase in the water level in front of the dam during flood seasons, the trap efficiency increases [8,9]. In the TGR, deposited sediment is mostly fine sediment with median diameters of less than 0.01 mm, which was not forecasted by previous studies [7]. In addition, fine-grained sediments with median sizes of approximately 0.004–0.010 mm have been observed in the dam area as early as the initial stage of the impoundment [10]. This sediment was deposited due to the significant increase in water depth of more than 100 m that occurred after the water storage operation. Although reservoir deposition is common, the sedimentation pattern in the dam area of the TGR is unusual. Generally, in the dam area, the distribution of sediment concentration is uniform, which would cause cross-sectional uniform deposition in the dam area, and then lead to an uneven surface across the section [11,12]. However, the deposits of fine-grained sediment in the dam area of the TGR have mainly accumulated in the deep channel and have a flat surface [13,14]. This has led to a rapid rise in the bed surface in the deep channel despite the small total amount of siltation. The elevations of the bottom discharge hole and deep discharge hole of the Three Gorges Dam (TGD) are 55 and 90 m, respectively. The bottom elevations of the water inlets of the power plant are 108 m. However, the surface elevation of the deposits in the deep channel before the TGD reached 60 m in September 2006, which was 20 m higher than that before the first impoundment in June 2003 [14]. Furthermore, the deposition has been persistently increasing. From 2003 to 2017, the total amount of accumulated sediment deposition in the near-dam area was 1.594×10^8 m^3; among them, the channel below 90 m elevation accounts for 73% of the total siltation. The average deposition thickness in the deep channel before the dam is 33.5 m, with a maximum deposition thickness of 63.7 m [15]. Of all the deposits, 0.47×10^8 m^3 was deposited during the period of 2008–2013, with a maximum thickness of 16.8 m [3]. Until recently, the elevation of the deposits was lower than that of the power plant inlets [16].

The rapid rise in the silt surface has greatly increased the probability of sediment entering the turbines. Hydraulic turbines may be subject to abrasion and corrosion by the heavier specific gravities of quartz and albite particles in sediment passing though, which will shorten its service life. On the other hand, the accumulation of sediment in the deep channel is conducive to the discharge of sediment from the reservoir. As the deposits have a certain fluidity (demonstrated below), the sediment can be discharged from the reservoir more easily by optimizing the dispatching measures [16].

The unique sedimentation pattern near the TGD requires a reasonable explanation based on scientific research. Models depicting the physical processes based on dynamic mechanisms are efficient research tools that would help to optimize dam operations and improve the theory of sediment transport in large reservoirs.

Many factors affect the cross-sectional deposition pattern in the reservoir, such as the sediment concentration, gradation, and lateral distributions of the flow velocity and water depth. The local river regime and cross-sectional shapes are also important. Han [17] ascribed the sedimentation pattern in which most deposits accumulate in the deep channel to the nonuniform crosswise distribution of sediment concentration and gradation. Jia et al. [18] proposed that the sedimentation pattern in the dam area of the TGR should be related to both the three-dimensional flow structure and the particle characteristics. During the early stage of reservoir impoundment, the medium size of suspended sediment before the dam was about 0.003–0.005 mm and the dry bulk density within the top 1 m of the deposits was about 0.4–1.0 t/m^3 [19]. Particles deposited in the dam area are so fine that their initial dry bulk density is small, and they are able to move as a fluid. Previous simulations of the sedimentation pattern in the dam area of the TGR did not consider the fluid characteristics of the deposits, and their predictions were significantly different from the observations [11,12,20]. The differences indicate the significance of considering the fluid characteristics of the deposits. By coupling a 1D gravity-driven flow model, which depicts the motion of sediment deposits, with a traditional 1D flow and sediment transport model, the accumulation processes of fine-grained sediments in the dam area of the TGR had been preliminarily identified [21,22]. However, more work is necessary to determine the mechanisms of the fine-grained sedimentation pattern in large reservoirs.

Considering the sedimentation features in the dam area of the TGR, we proposed a physical-based numerical model for fine-grained sediment deposition in large reservoirs. A method of modeling fine-grained sediment movement that includes gravity-driven motion in reservoirs was developed and applied, along with a 3D Reynolds-averaged numerical model of water flow and sediment transport, to simulate the uncommon sedimentation pattern in the dam area of the TGR. Our method can be applied to other reservoirs with similar sedimentation problems.

2. Materials and Methods

2.1. Features of Fine-Grained Sediment Deposition in Deep Reservoirs

The Three Gorges Dam (TGD), which is located in the Three Gorges reach along the trunk of the Yangtze River in China, is a key backbone project for flood control, navigation, and water resources development and has a total installed capacity of 18.2 GW and an average annual output of 84.7 billion kWh. The dam is located in Sandouping, 40 km upstream from Yichang City, as shown in Figure 1. The normal pool level of the Three Gorges Reservoir (TGR) is 175 m (relative to the Chinese water level zero point), which corresponds to an aggregated storage capacity of 39.3 billion m^3. The TGR has been in operation since June 2003. The construction of large hydraulic projects such as the TGD has generated problems in navigation, environment, and ecology, all of which are related to sediment transport processes, including advection, diffusion, deposition, and erosion.

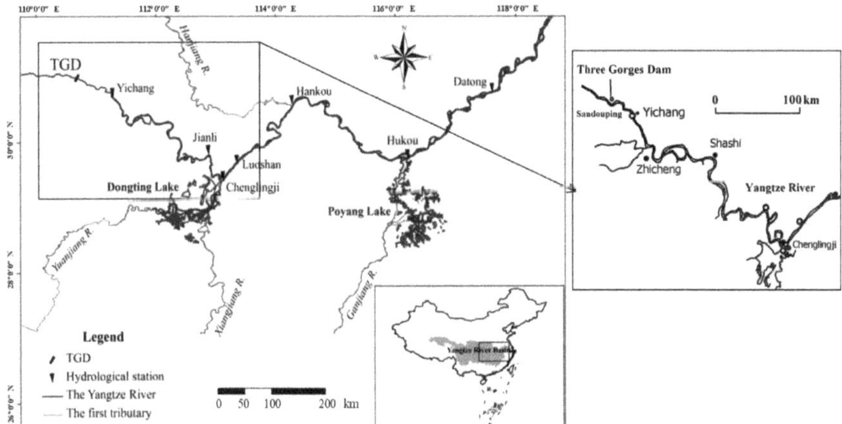

Figure 1. Location of the TGD on the Yangtze River. (The left part of the figure is based on [23]).

Observations of the sedimentation pattern in the dam area of a typical section (S34#, see Figure 2 for the location) of the TGR have shown that the sediment mainly accumulated in the deep channel with a nearly horizontal surface (Figure 3) in the early years of reservoir operation. Conversely, deposits of fine-grained sediment in reservoirs typically have a relatively uniform distribution across the section. The physical and dynamic characteristics of the sediment deposits in the dam area can be used to understand this difference.

Before impoundment, the median size of the riverbed materials near the dam was approximately 0.10 mm. With the initial impoundment of the TGR, the water level rose to 135 m, and large amounts of siltation occurred in the river reach before the dam. In October 2003, the observed median particle size of the bed materials had decreased significantly to 0.02 mm. With further deposition, the median size had further decreased to 0.008 mm in May 2006. After the 156 m water storage operation of the TGR, the observed median size of the bed materials in the dam area was 0.005 mm in December 2006 and October 2007 [10]. The silt particles deposited in the dam area are extremely fine and require a long time to compact. Before compaction, the low-density deposits (the bulk density is approximately 1.00–1.20 t/m^3) in the form of sludge are called fluid mud. The fluid mud has significant fluidity and can flow or deform under its own weight. If the surface slope exceeds a certain value, the fluid mud will move down the slope. This may be the main reason that most of the deposits were found in the deep channel of the TGR and that the surface of the sludge is horizontal.

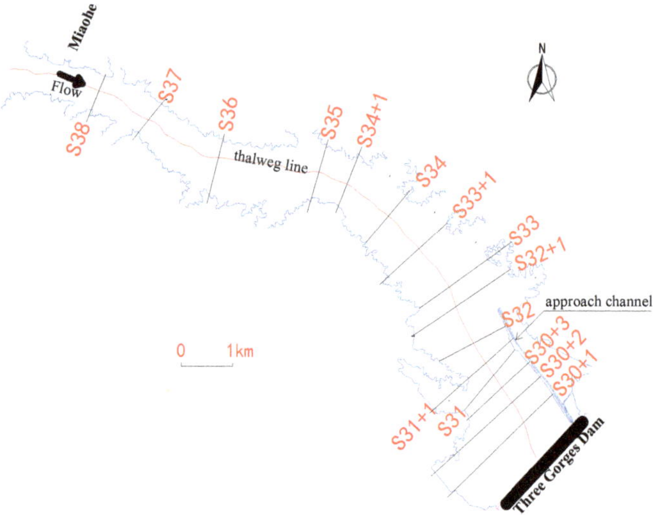

Figure 2. Sketch of the dam area in the TGR.

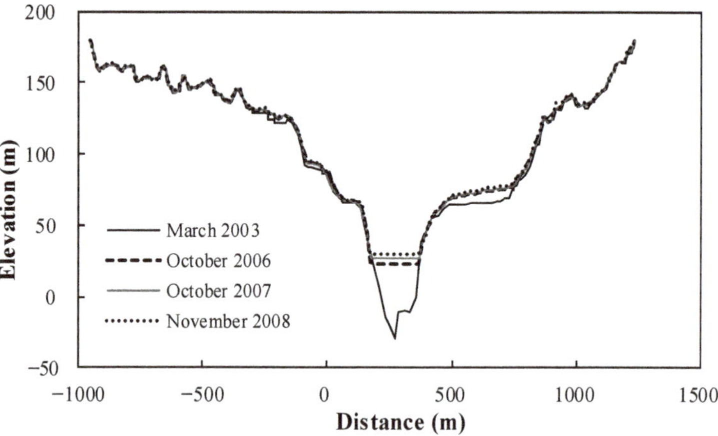

Figure 3. Observed sedimentation pattern on a typical cross-section (Section S34#) near the TGD.

Therefore, the sedimentation process of fine sediment in deep areas of reservoirs can be summarized as several physical processes, including the dynamic process of sediment transport and settlement in flowing water, gravity-driven fluid flow of deposits, and compaction and consolidation.

The gravity-driven fluid flow of fine deposits is similar to that of fluid mud. Suppose that the surface slope of the fluid mud layer is J_c. If $J_c > \tau_B / (\gamma_s \Delta z)$ (where τ_B is the Bingham ultimate shear force, and γ_s and Δz are the weight and thickness of the fluid mud layer, respectively), the fluid mud would move down the slope under its own weight, as shown in Figure 4. Field observations have shown that the deposits in the deep channel are thick and that their surface is nearly horizontal in the dam area of the TGR (as shown in Figure 3). The main reason for this phenomenon is that the suspended sediment that

was deposited during the early period of the TGR impoundment was so fine (median size of 0.004–0.01 mm) that the resulting deposits had a low initial dry bulk density and could move like fluid mud. The mud-like deposits moved downward of the surface slope due to gravity and accumulated in the deep channel of the riverbed, forming an uncommon sedimentation pattern in the early stage of the impoundment of the large reservoir.

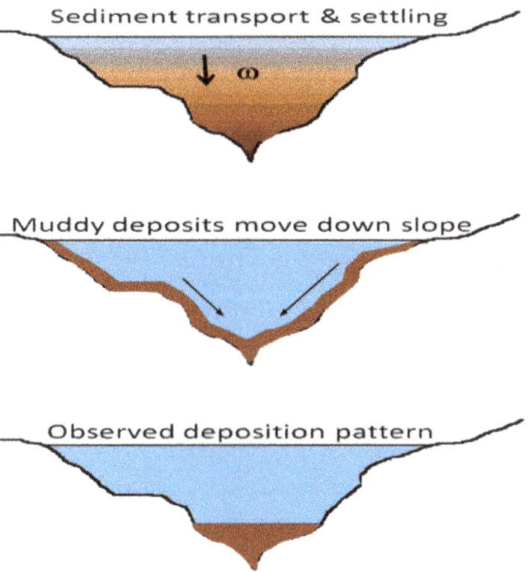

Figure 4. Schematic processes of fine-grained particle sedimentation in the dam area [20].

2.2. Numerical Simulation Method for Transport and Settlement of Fine-Grained Sediment

Based on the analysis presented above, the whole processes of fine-grained sediment in the deep reservoir can be summarized as transport and settlement, gravity-driven fluid flow, and consolidation. The 3D Reynolds-averaged numerical model that depicts flow, sediment transport, and sediment settlement in a reservoir can be found in the related literature [20]. Here, we briefly describe the method.

The governing equations for water flow are given as:

$$\frac{\partial u_i}{\partial x_i} = 0 \tag{1}$$

$$\frac{\partial u_i}{\partial t} + \frac{\partial (u_i u_j)}{\partial x_j} = -\frac{1}{\rho}\frac{\partial p}{\partial x_i} + \frac{1}{\rho}\frac{\partial \tau_{ij}}{\partial x_j} + f_i \tag{2}$$

where u_i ($i = 1,2,3$) are velocity components; f_i is the gravitational force per unit volume; ρ is the density of water; and p is the pressure. In this model, turbulence stresses τ_{ij} are calculated with the standard $k - \varepsilon$ turbulence model.

The 3D nonuniform suspended sediment transport equation can be expressed in Cartesian coordinates as:

$$\frac{\partial s_k}{\partial t} + \frac{\partial}{\partial x}(u s_k) + \frac{\partial}{\partial y}(v s_k) + \frac{\partial}{\partial z}(w s_k) = \frac{\partial}{\partial x}\left(\varepsilon_s \frac{\partial s_k}{\partial x}\right) + \frac{\partial}{\partial y}\left(\varepsilon_s \frac{\partial s_k}{\partial y}\right) + \frac{\partial}{\partial z}\left(\varepsilon_s \frac{\partial s_k}{\partial z}\right) + \frac{\partial}{\partial z}(\omega_{sk} s_k) \tag{3}$$

where u, v, and w are the flow velocities in the x, y, and z directions, respectively; s_k is the sediment concentration of group k; ω_{sk} is the settling velocity of group k; and ε_s is the sediment diffusion coefficient.

The equation of bed evolution caused by suspended sediment transport (deposition) is

$$\gamma'_s \frac{\partial Z_b}{\partial t} = \sum_{1}^{N_s} \omega_{sk}(s_{kb} - s_{kb*}) \tag{4}$$

where γ'_s is the dry bulk density of the deposits; s_{kb} and s_{kb*} are the real and equilibrium sediment concentrations of group k near the bed, respectively; Z_b is the bed level; and N_s is the total number of sediment groups.

2.3. Numerical Simulation Method for Gravity-Driven Motion of Fine-Grained Sediment Deposits

Solid consolidation can be determined using existing methods. Here, we mainly discuss the method of simulating the gravity-driven flow of the fine-grained sediment deposits. Newly deposited fine-grained sediment deposits are a non-Newtonian fluid and are generally considered to be a Bingham fluid. Before movement, it is necessary to overcome the Bingham ultimate shear force. In this paper, a critical slope is used as a criterion for the destabilizing flow of fine-grained sediment deposits. The shallow water flow model is used to describe the gravity-driven flow process of the destabilized deposits.

When the surface slope of fine-grained sediment deposits J_c satisfies the condition $J_c > \tau_B/(\gamma_s \Delta z)$ (where τ_B is the Bingham ultimate shear force, and γ_s and Δz are the weight and thickness of the fluid mud layer, respectively), the deposits will move down the slope under their own weight [24]. Fine-grained deposits with different bulk densities have different Bingham ultimate shear forces. Here, we apply the formula for the Bingham ultimate shear force on fluid mud on a muddy shore [25],

$$\tau_B = \alpha \times \exp(\beta \times \gamma_s) \tag{5}$$

where α and β are empirical parameters; $\alpha = 9 \times 10^{-9}$; and $\beta = 16.719$ based on previous research [25].

After losing stability, the fine-grained deposit will flow. As the thickness of the deposits is small compared to the horizontal scale of the reservoir, we assume that the flow of the fine-grained deposit can be described by the 2D shallow flow equations.

Continuity equation:

$$\frac{\partial Z}{\partial t} + \frac{\partial M}{\partial x} + \frac{\partial N}{\partial y} = q \tag{6}$$

Momentum equation:

$$\frac{\partial M}{\partial t} + \frac{\partial u M}{\partial x} + \frac{\partial v M}{\partial y} = -gh\frac{\partial Z}{\partial x} + D\left(\frac{\partial^2 M}{\partial x^2} + \frac{\partial^2 M}{\partial y^2}\right) - \frac{gn^2 M\sqrt{u^2 + v^2}}{h^{\frac{4}{3}}} + qu_0 \tag{7}$$

$$\frac{\partial N}{\partial t} + \frac{\partial u N}{\partial x} + \frac{\partial v N}{\partial y} = -gh\frac{\partial Z}{\partial y} + D\left(\frac{\partial^2 N}{\partial x^2} + \frac{\partial^2 N}{\partial y^2}\right) - \frac{gn^2 N\sqrt{u^2 + v^2}}{h^{\frac{4}{3}}} + qv_0 \tag{8}$$

where Z is the surface elevation of the fine-grained deposits; $M = uh$; $N = vh$; h is the thickness of the deposits; u and v are the velocities in the x and y directions, respectively; q is the source/sink of the fine deposit in a unit area, which can be calculated according to the instability as $q = (\Delta z - \Delta z_c)/dt$, where Δz_c is the maximum thickness that can remain stable, and $\Delta z_c = \frac{\tau_B}{(J_c \gamma_s)}$; g is the acceleration due to gravity; D is the kinematic viscosity; and n is the integrated Manning's roughness. As gravity is the main driving force of the flow, the diffusion terms in Equations (7) and (8) can be neglected.

2.4. Numerical Simulation Procedure

Consistent with the 3D flow and sediment transport model, the gravity-driven flow equations of fine deposits are discretized based on the finite volume method, which has the advantage of conservation. The SIMPLEC algorithm is used to solve the problem. The solution process is shown in Figure 5 and illustrated as follows.

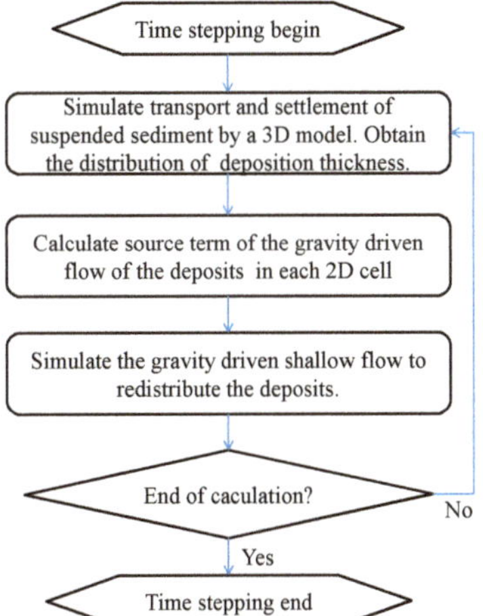

Figure 5. Numerical simulation procedure.

Step (1): Use the 3D numerical model to simulate the flow, sediment transport, and sediment deposition and to obtain the distribution of the deposition thickness Δz.

Step (2): Based on the silt thickness and terrain slope, determine whether the deposits in a cell are stable or not. If $J_c > \tau_B / (\gamma_s \Delta z)$, the deposit is unstable and the source term q will be calculated; otherwise, the deposit is stable and q is zero.

Step (3): Solve the deposit flow equations and apply the SIMPLEC algorithm to obtain the surface elevation of the deposits after being redistributed.

Step (4): Repeat steps (1)–(3) until the end of the calculation.

3. Results

Based on the topographic, hydrological, and sediment data measured in the initial stage of the TGR impoundment, a 3D Reynolds-averaged water flow and sediment transport numerical model in the dam area of the TGR (from Miaohe to the dam, see Figure 2) was established. The simulated area was approximately 15 km long. The initial terrain for the simulation was measured in June 2003 (Figure 6). Flow and sediment data at Miaohe from June 2003 to November 2008 were used for the inflow conditions. The water level in front of the dam was used for the outflow conditions. The numbers of grid points in the longitudinal, lateral, and vertical directions were 198, 97, and 18, respectively.

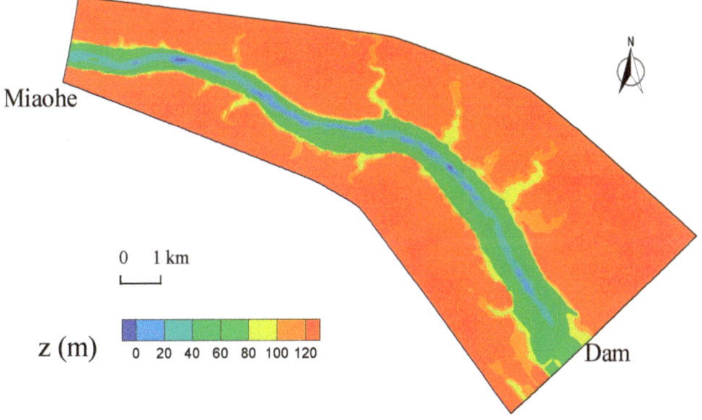

Figure 6. Observed bed elevation near the Three Gorges Dam (June 2003).

3.1. Flow Field and Suspended Sediment Concentration

Figure 7 shows the calculated flow field for a flow discharge of 16,700 m^3/s and a pool level of 135 m above sea level. Strong 3-D flow features can be observed in the bends near the dam area, and secondary currents are significant. The velocity vectors near the bottom point to the convex bank, whereas the surface vectors point to the concave bank. Based on the calculated flow velocities, the suspended load concentration can be calculated by solving the convection–diffusion equation. Figure 8 shows the calculated suspended load concentrations on typical cross-sections. The sediment concentration decreases much more rapidly in wider valley areas. Due to the impact of the secondary currents shown in Figure 7, the maximum concentration occurs at the convex side, whereas the minimum occurs at the concave side. Figure 9 shows a comparison of the observed and simulated velocities and suspended load concentrations on typical cross-sections (see Figure 2 for the locations of the sections) for a flow discharge of 24,600 m^3/s. In general, the 3D numerical model accurately predicted the flow patterns and suspended load concentrations in the dam area of the TGR.

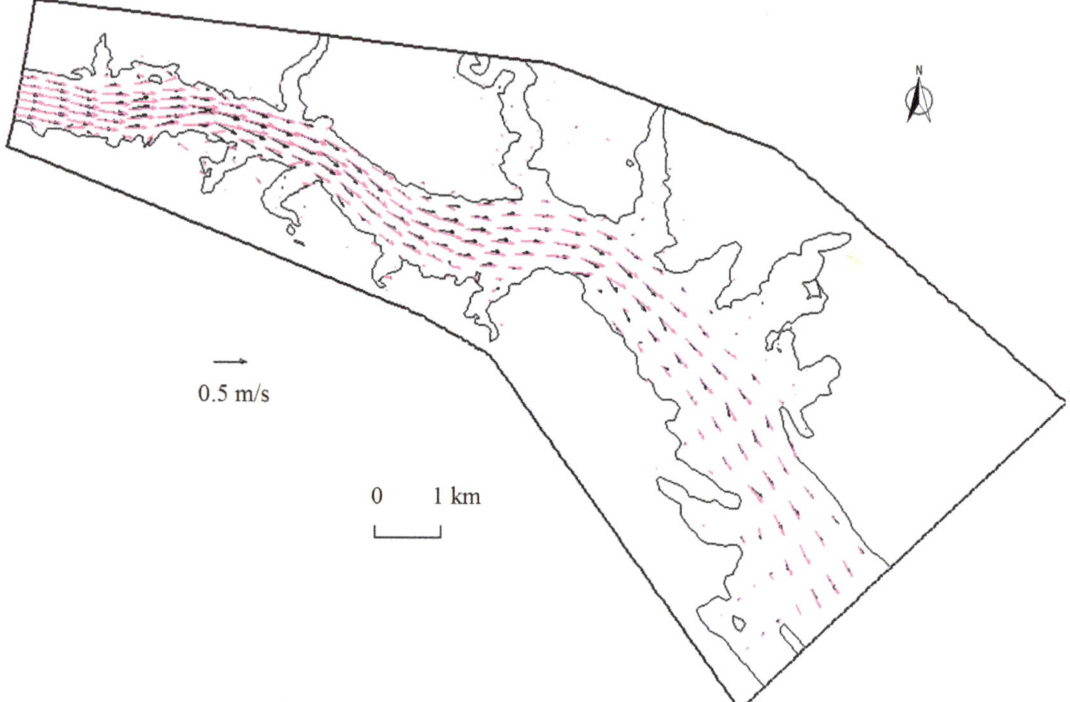

Figure 7. Calculated velocity fields near the bottom and the water surface (the blue lines are the velocity vectors close to the water surface, and the black lines are the vectors close to the bottom).

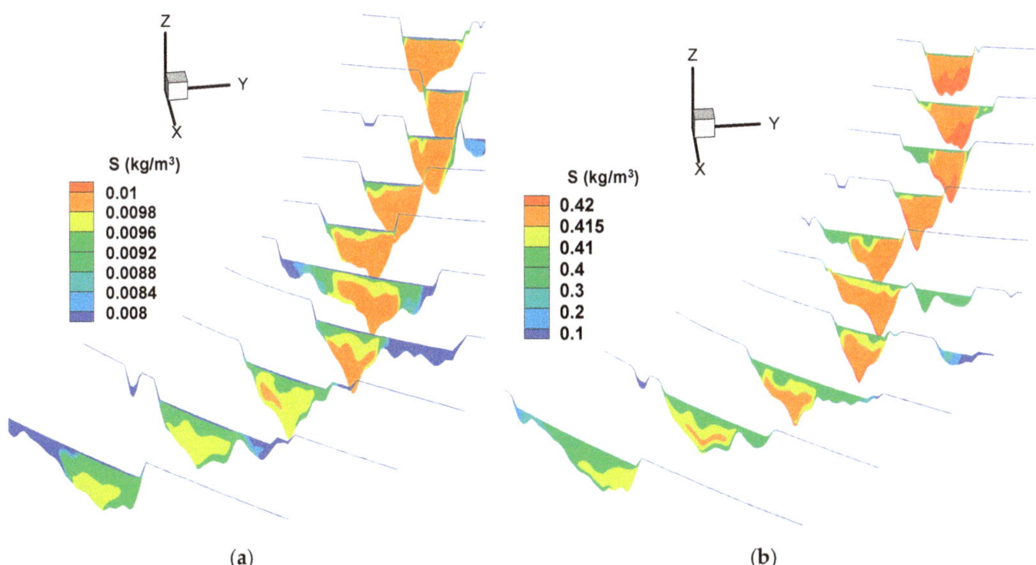

Figure 8. Simulated suspended load concentrations on typical cross-sections. (**a**) the discharge is 6300 m^3/s and water level is 139 m; (**b**) the discharge is 32,100 m^3/s and water level is 135 m.

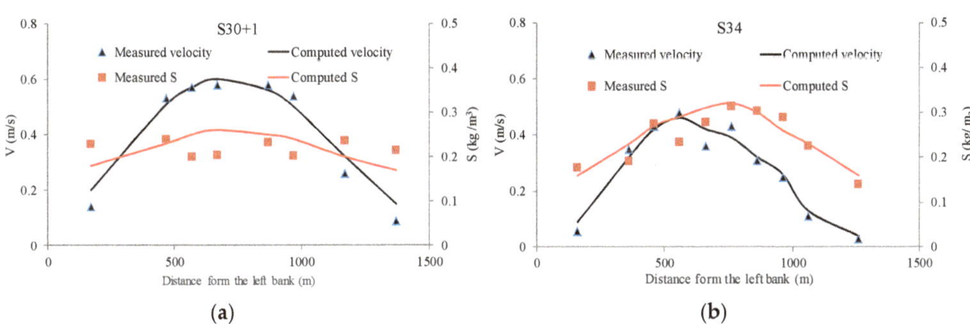

Figure 9. Verification of the velocity and sediment concentration distributions on typical cross-sections. (**a**) on section S30+1; (**b**) on section S34.

3.2. Features of Sedimentation Pattern

The river reach in front of the TGD is a wide valley of crystalline rocky hills with significant slopes on both banks. Fine-grained sediment settled in this area after the impoundment of the TGR. If the thickness of the deposits in a certain location exceeds a critical value, the deposits will flow down the slope under their own weight. Based on the verifications of the flow velocity and sediment concentration (Figure 9), Figure 10 shows the sediment thickness near the dam calculated by traditional 3D flow and sediment transport modeling (without considering the gravity-driven fluid flow of the deposits), and Figure 11 shows the results calculated using the proposed method. Although the total amounts of sediments are consistent, the sediment distributions are quite different. The sediment thickness calculated by the traditional method, which does not consider the fluid flow of the fine-grained deposit, is relatively uniform in map view (Figure 10). In contrast, the proposed model (Figure 11) predicts that most of the deposits will accumulate in the deep channel and that a few deposits will remain on the flat terraces on both sides, which is consistent with the observations.

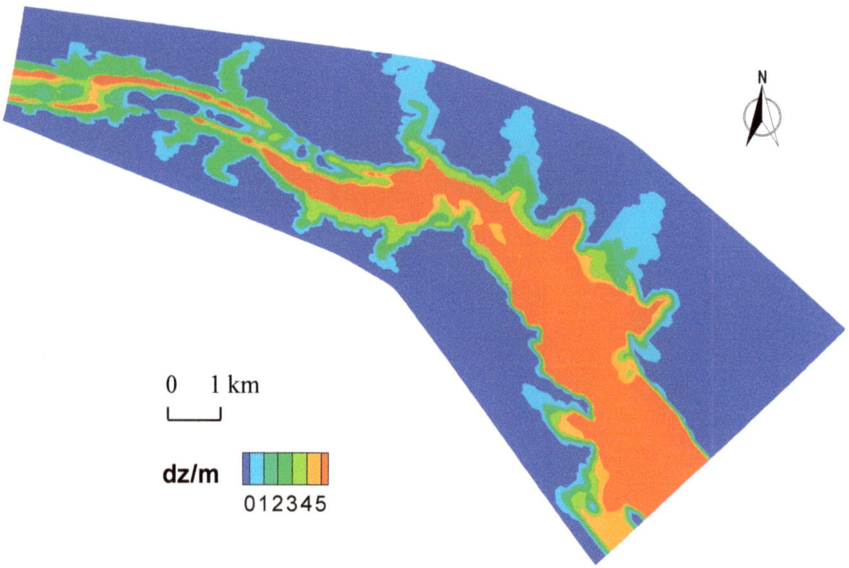

Figure 10. Computed sediment distribution using the traditional method.

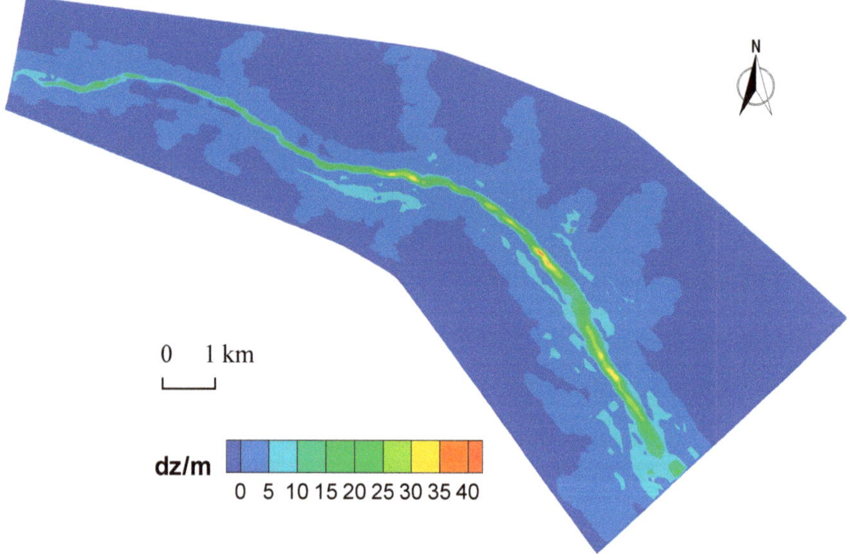

Figure 11. Computed sediment distribution considering gravity-driven processes.

Figure 12 shows comparisons between the observed and calculated sedimentation on several typical cross-sections (section locations are shown in Figure 2). The cross-sectional sedimentation pattern that considers gravity-driven motion is more similar to the observations than the pattern that does not consider the motion. This again confirms the significance of the gravity-driven flow of fine deposits.

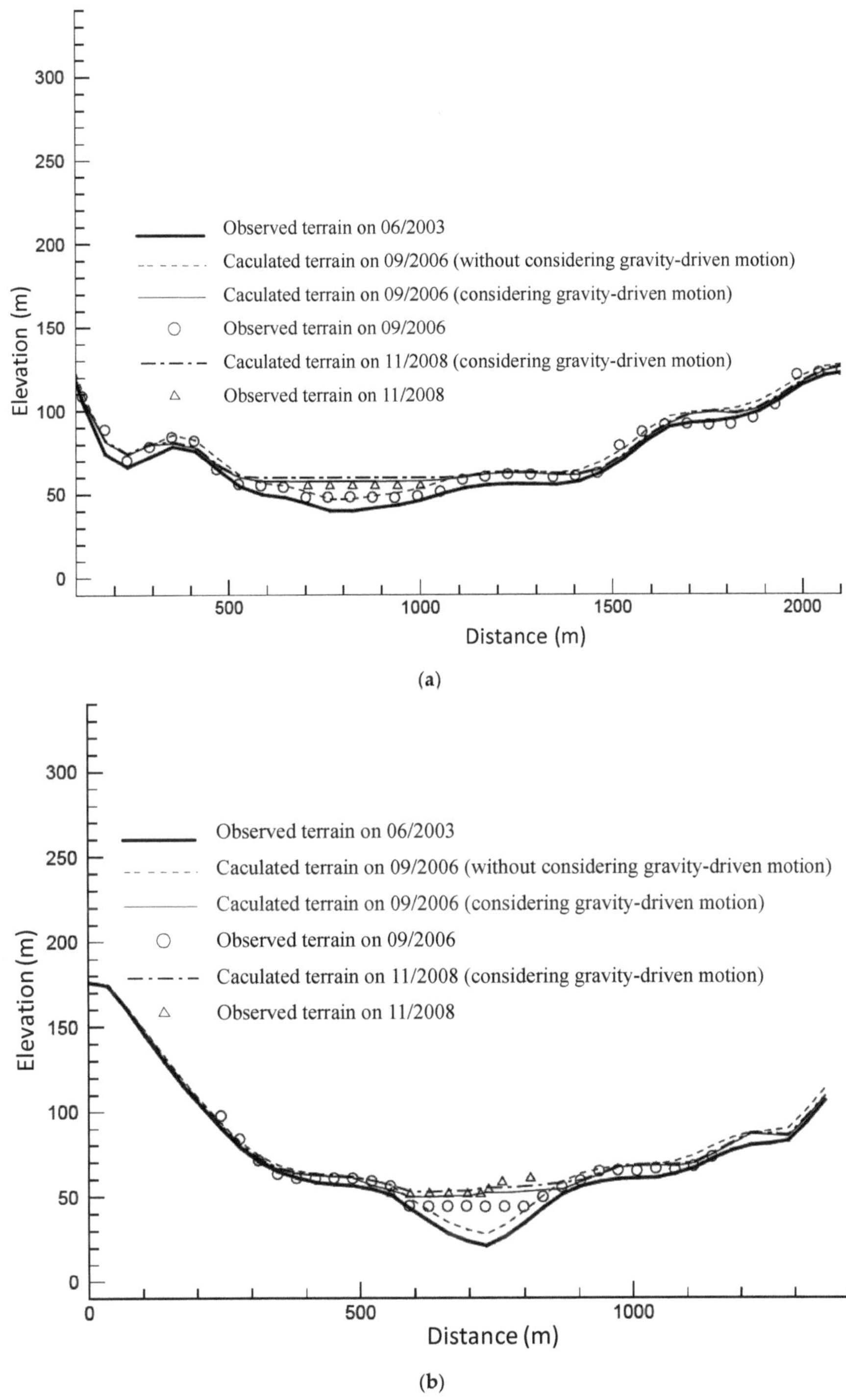

Figure 12. *Cont.*

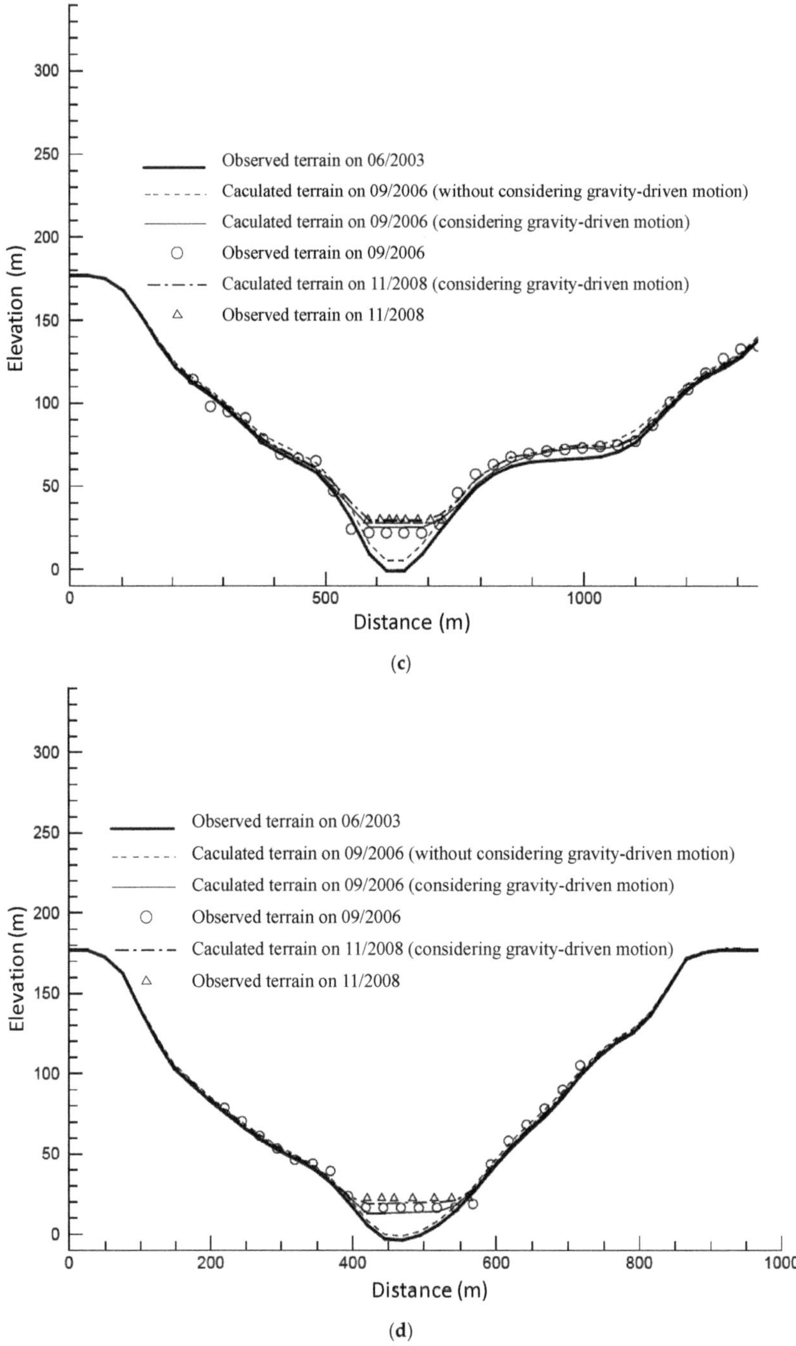

Figure 12. Comparisons of the sediment distributions on a typical cross-section between the observations and simulated results. (**a**) on section S30+1; (**b**) on section S32; (**c**) on section S34; (**d**) on section S37.

Figure 13 compares the surface elevations along the thalweg between the observations and the results calculated with and without considering the gravity-driven deposit flow. It can be seen that the calculated results that consider the deposit flow match the observations quite well. Except the point at the entrance, the maximum error at the other points is 5.1 m, which is located at section S31# (see Figure 2 for the location). Based on measured data, the averaged and maximum deposition thickness along the thalweg are 22.22 and 57.3 m, respectively. Based on simulating results without considering the gravity-driven flow, the two values are 8.32 and 31.5 m correspondingly. The corresponding values based on simulating results that consider the gravity-driven flow are 24.43 and 56.5 m. By considering the gravity-driven flow, the averaged error of deposition thickness along the thalweg decreases from −13.9 to 2.2 m.

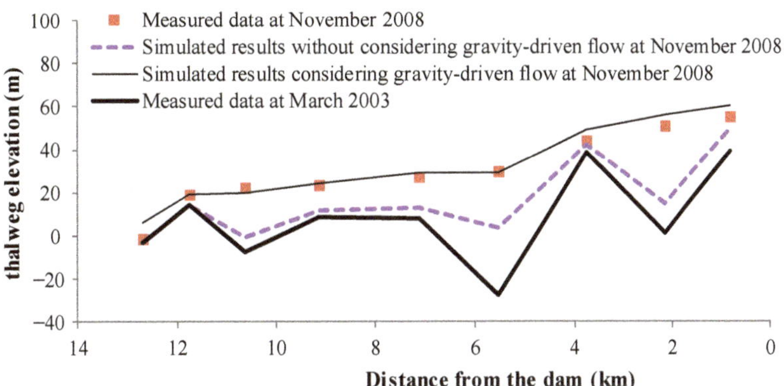

Figure 13. Computed and observed elevations of the thalweg.

4. Discussion

4.1. Difference between Calculation and Measurement

Although the comparison strongly confirmed the necessity and advantage of considering the gravity-driven movement of the deposits, there was some discrepancy between the observations and simulating results.

Near the dam, as shown in Figures 12a and 13, the depth of the deposit at the thalweg is greater than that of the observation. This is probably because this section is near the discharge hole, so the fluid deposits may be scoured and flushed by the flow. The effect of discharge flow was not simulated precisely in this study, which may have caused the deviation. Figure 14 shows the computed velocity distribution in front of the dam. In this work, the outlets of the power plant were generalized as a single outlet area, which would flatten the flow speed nearby. Another possible reason is the terrain flattening caused by mesh interpolation in the numerical simulation. Without terrain flattening, some of the deposit may be blocked by local terrain undulation, reducing the amount of deposits that gathered in the deep channel. It is easy to imagine that those cross-sections with wider sides, more undulating, relatively gentle side slopes would suffer greater impact from terrain flattening. As the reach of the near-dam area gradually widens from Miaohe to the dam, the terrain flattening error is generally more obvious at the sections before the dam, which can explain the relatively larger discrepancy within 5 km before the dam than that of the upstream results in Figure 13.

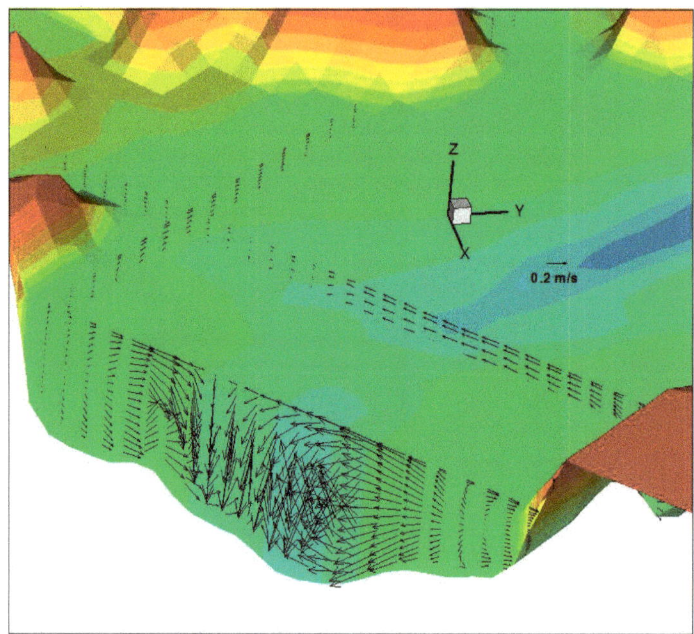

Figure 14. Computed velocity distribution in front of the dam (the discharge is 35,600 m^3/s and water level is 135.1 m).

4.2. Shortcomings of Present Method

In this work, shallow water equations are applicable to a Newtonian fluid. For a Bingham fluid, when integrating the continuity and momentum equations with depth to obtain the 2D governing equations, the nonlinear relationship between the shear stress and strain would generate significant differences with the shallow water equations [26]. If the shallow water equations are used, the resistance term would be more complicated than that of Manning's formula [27]. Therefore, the modeling of gravity-driven flow of fine-grained deposits should be improved further. In addition, due to the discretization, the topographic data used in the numerical simulation are not consistent with the real topography. This also contributed to the discrepancy between the simulated results and observations. However, considering the gravity-driven fluid flow of the deposits, the simulation results of the overall sedimentation trend and distribution showed good agreements with the observations.

5. Conclusions

The deposition processes of fine-grained sediment in large reservoirs can be summarized as the settling of suspended sediment, gravity-driven fluid flow of the deposits, and sludge consolidation and compaction. Based on the physical mechanisms, we analyzed the dynamic causes of the patterns of fine-grained sedimentation in the dam area of the TGR during the initial stage of impoundment. The sediment deposited in the dam area of the TGR during this stage is mostly fine-grained sediment with a grain size generally between 0.004 and 0.010 mm. These fine-grained sediment deposits have a low dry bulk density and can move downslope toward the bottom of the riverbed due to the effect of gravity. These fluid deposits finally gather in the deep channel and have a horizontal surface, which is an uncommon feature of sedimentation in large reservoirs. The rapid increase in the surface elevation of the silt in front of the dam threatens the operations of the dam.

By coupling a traditional 3D water flow and sediment transport model with a newly proposed numerical method for the gravity-driven flow of fine-grained sediment deposits, a new method was developed and used to simulate the sedimentation pattern in the dam area of the TGR. The results showed that sediment would be deposited relatively uniformly in the dam area without considering the gravity-driven flow. When the gravity-driven fluid flow of the fine-grained deposits is considered, the distribution of the sediment deposits was more consistent with the observations than the results that do not consider gravity-driven flow. The simulation results showed that the deposits mainly accumulated in the deep channel with a flat surface, which agreed well with the observations. This study revealed the key role of the gravity-driven flow of deposits in simulating the deposition of fine-grained sediment in large reservoirs.

Author Contributions: Conceptualization, D.J. and X.S.; funding acquisition, D.J. and X.Z.; investigation, D.J. and X.Z.; methodology, D.J. and J.Z.; project administration, D.J. and X.Z.; resources, D.J. and X.Z.; software, D.J.; supervision, X.S.; validation, D.J. and J.Z.; writing—original draft, J.Z.; writing—review and editing, D.J. All authors have read and agreed to the published version of the manuscript.

Funding: This research was funded by the National Key R&D Program of China (Nos. 2018YFC0407402) and the National Natural Science Foundation of China (Nos. 51579151, 51779148 and U2040215).

Institutional Review Board Statement: Not applicable.

Informed Consent Statement: Not applicable.

Data Availability Statement: No new data were created or analyzed in this study. Data sharing is not applicable to this article.

Conflicts of Interest: The authors declare no conflict of interest.

References

1. Zhou, J.J.; Zhang, M.; Cao, H.Q Removing coarse sediment by sorting of reservoirs. *Sci. China Technol. Sci.* **2011**, *41*, 833–844. (In Chinese) [CrossRef]
2. Schleiss, A.J.; Franca, M.J.; Juez, C.; De Cesare, G. Reservoir sedimentation. *J. Hydraul. Res.* **2016**, *54*, 595–614. [CrossRef]
3. Huang, Y.; Wang, J.; Yang, M. Unexpected sedimentation patterns upstream and downstream of the Three Gorges Reservoir: Future risks. *Int. J. Sediment. Res.* **2019**, *34*, 108–117. [CrossRef]
4. Faghihirad, S.; Lin, B.; Falconer, R. Application of a 3D Layer Integrated Numerical Model of Flow and Sediment Transport Processes to a Reservoir. *Water* **2015**, *7*, 5239–5257. [CrossRef]
5. The Science and Technology Department of Ministry of Water Conservancy and Electric Power. *Reports Collection on the Sediment Issues of the Three Gorges Reservoir*; The Science and Technology Department of Ministry of Water Conservancy and Electric Power: Beijing, China, 1988. (In Chinese)
6. State Council Three Gorges Project Construction Committee Executive Office Sediment Experts Group, China Three Gorges Project Corporation Sediment Experts Group. *Sediment Research of the Three Gorges Project (1996–2000, Volume 8)*; Intellectual Property Press: Beijing, China, 2008; Chapter 3. (In Chinese)
7. Li, W.J.; Yang, S.F.; Fu, X.H.; Xiao, Y. Study on the sedimentation characteristics in the Three Gorges Reservoir during the initial operation stage. *Adv. Water Sci.* **2015**, *26*, 676–685. (In Chinese)
8. Chen, G.Y.; Yuan, J.; Xu, Q.X. On sediment diversion ratio after the impoundment of the Three Gorges Project. *Adv. Water Sci.* **2012**, *23*, 355–362. (In Chinese)
9. Yang, C.R.; Deng, J.Y.; Qi, Y.M.; Wang, Y.X.; Lin, J.Y. Study on Long-term Impact of Small and medium-sized floods regulation on sedimentation of the Three Gorges Reservoir. *Water Resour. Power* **2020**, *38*, 34–37. (In Chinese)
10. Bureau of Hydrology, Changjiang Water Resources Commission. *Analysis of Sedimentation and Erosion in the Three Gorges Reservoir and the River Reach Between TGP and Gezhouba Project in 2007*; Bureau of Hydrology, Changjiang Water Resources Commission: Wuhan, China, 2008. (In Chinese)
11. Fang, H.W.; Rodi, W. Three-dimensional calculations of flow and suspended sediment transport in the neighborhood of the dam for the Three Gorges Project reservoir in the Yangtze River. *J. Hydraul. Res.* **2003**, *41*, 379–394. [CrossRef]
12. Lu, Y.J.; Dou, G.R.; Han, L.X.; Shao, X.J.; Yang, X.H. 3D mathematical model for suspended load transport by turbulent flows and its applications. *Sci. China. Ser. E.* **2004**, *47*, 237–256. [CrossRef]
13. Zhang, N.; Yi, D.R.; Dai, W.L. Analysis of scouring and silting in the reach from dam to Lidu town in the Three Gorges Reservoir. *Yangtze River* **2006**, *37*, 95–98. (In Chinese)
14. Department of Hydraulic Engineering, Tsinghua University. *Analysis of the Field Data of Reservoir Sedimentation in the Three Gorges Reservoir*; Tsinghua University: Beijing, China, 2009. (In Chinese)
15. Hu, C. Analysis on sediment scouring and silting variation of Three Gorges Reservoir since 175 m trial impoundment for past ten years. *Water Resour. Hydropower Eng.* **2019**, *50*, 18–26. (In Chinese)

16. Ren, S.; Zhang, B.; Wang, W.; Yuan, Y.; Guo, C. Sedimentation and its response to management strategies of the Three Gorges Reservoir, Yangtze River, China. *Catena* **2021**, *199*, 105096. [CrossRef]
17. Han, Q.W. *Reservoir Sedimentation*; Science Press: Beijing, China, 2003. (In Chinese)
18. Jia, D.D.; Shao, X.J.; Zhang, X.N.; Zhou, J.Y. Preliminary analysis on the causes of reservoir sedimentation pattern in the vicinity of Three Gorges Project during its early filling. *Adv. Water Sci.* **2011**, *22*, 539–545. (In Chinese)
19. Zhang, N.; Dai, W.; Zhu, H.; Li, J. Initial dry density of sediments in the Three Gorges Reservoir. *Yangtze River* **2006**, *37*, 59–61. (In Chinese)
20. Jia, D.D.; Shao, X.J.; Zhang, X.N.; Ye, Y.T. Sedimentation patterns of fine-grained particles in the dam area of the Three Gorges Project: 3D numerical simulation. *J. Hydraul. Eng-ASCE* **2013**, *139*, 669–674. [CrossRef]
21. Zhou, J.Y.; Shao, X.J.; Jiang, L.; Jia, D.D. Gravity-driven transport of fine grained reservoir sediments. *J. Zhejiang Univ.* **2014**, *48*, 2254–2258. (In Chinese)
22. Xiao, Y.; Yang, F.S.; Zhou, J.Y.; Chen, W.S. 1-D numerical modeling of the mechanics of gravity-driven transport of fine sediments in the Three Gorges Reservoir. *Lake. Reserv. Manag.* **2015**, *31*, 83–91. [CrossRef]
23. Jiang, L.; Ban, X.; Wang, X.; Cai, X. Assessment of Hydrologic Alterations Caused by the Three Gorges Dam in the Middle and Lower Reaches of Yangtze River, China. *Water* **2014**, *6*, 1419–1434. [CrossRef]
24. Chien, N.; Zhang, R.; Zhou, Z.D. *Fluvial Processes*; Science Press: Beijing, China, 1987. (In Chinese)
25. Tang, L.; Zhang, W.; Wu, F.L. Study on sediment movement forms and mechanism causing mild slope on nearshore muddy coast. *J. Sediment Res.* **2016**, *6*, 66–73. (In Chinese)
26. Imran, J.; Parker, G.; Locat, J.; Lee, H. 1D numerical model of muddy subaqueous and subaerial debris flows. *J. Hydraul. Eng.* **2001**, *127*, 959–968. [CrossRef]
27. Laigle, D.; Coussot, P. Numerical modeling of mud flows. *J. Hydraul. Eng.* **1997**, *123*, 617–623. [CrossRef]

Article

Sediment Balance Estimation of the 'Cuvette Centrale' of the Congo River Basin Using the SWAT Hydrological Model

Pankyes Datok [1,*], Sabine Sauvage [1,*], Clément Fabre [2], Alain Laraque [3], Sylvain Ouillon [4,5],
Guy Moukandi N'kaya [6] and José-Miguel Sanchez-Perez [1]

[1] Laboratoire Ecologie Fonctionnelle et Environnement, Université de Toulouse, CNRS, INPT, UPS,
    31326 Toulouse, France; jose-miguel.sanchez-perez@univ-tlse3.fr
[2] Eawag, Swiss Federal Institute of Aquatic Science and Technology, CH-8600 Dübendorf, Switzerland;
    clement.fabre21@gmail.com
[3] GET, UMR CNRS/IRD/UPS, 31400 Toulouse, France; alain.laraque@ird.fr
[4] LEGOS, Université de Toulouse, IRD, CNRS, CNES, UPS, 31400 Toulouse, France;
    sylvain.ouillon@legos.obs-mip.fr
[5] USTH, Vietnam Academy of Science and Technology (VAST), Hanoi 100000, Vietnam
[6] LMEI/CUSI/ENSP/Université Marien Ngouabi, Brazzaville BP 69, Congo; guymoukandi@yahoo.fr
*   Correspondence: pankyes-emmanuel.datok@univ-tlse3.fr (P.D.); sabine.sauvage@univ-tlse3.fr (S.S.)

Citation: Datok, P.; Sauvage, S.;
Fabre, C.; Laraque, A.; Ouillon, S.;
Moukandi N'kaya, G.; Sanchez-Perez,
J.-M. Sediment Balance Estimation of
the 'Cuvette Centrale' of the Congo
River Basin Using the SWAT
Hydrological Model. *Water* **2021**, *13*,
1388. https://doi.org/
10.3390/w13101388

Academic Editor: Bommanna
Krishnappan

Received: 20 April 2021
Accepted: 14 May 2021
Published: 16 May 2021

Abstract: In this study, the SWAT hydrological model was used to estimate the sediment yields in the principal drainage basins of the Congo River Basin. The model was run for the 2000–2012 period and calibrated using measured values obtained at the basins principal gauging station that controls 98% of the basin area. Sediment yield rates of 4.01, 5.91, 7.88 and 8.68 t km^{-2} yr^{-1} were estimated for the areas upstream of the Ubangi at Bangui, Sangha at Ouesso, Lualaba at Kisangani, and Kasai at Kuto-Moke, respectively—the first three of which supply the Cuvette Centrale. The loads contributed into the Cuvette Centrale by eight tributaries were estimated to be worth 0.04, 0.07, 0.09, 0.18, 0.94, 1.50, 1.60, and 26.98 × 10^6 t yr^{-1} from the Likouala Mossaka at Makoua, Likouala aux Herbes at Botouali, Kouyou at Linnegue, Alima at Tchikapika, Sangha at Ouesso, Ubangi at Mongoumba, Ruki at Bokuma and Congo at Mbandaka, respectively. The upper Congo supplies up to 85% of the fluxes in the Cuvette Centrale, with the Ubangi and the Ruki contributing approximately 5% each. The Cuvette Centrale acts like a big sink trapping up to 23 megatons of sediment produced upstream (75%) annually.

Keywords: Congo River Basin; Cuvette Centrale; erosion; sediment transport; sediment balance; SWAT

1. Introduction

The most important agent transporting sediment across the land is river transport, while relief and runoff are the most important factors determining the sediment load of rivers [1] (Hay, 1998). The suspended sediment load transported by a river or stream is a mixture of sediment derived from different areas in a catchment that comprise source types with different erosional processes depending on the time and place [2,3] (Walling, 2006; Haddadchi et al., 2013). Sediment fluxes and their variability are important and merit study because drastic changes in their amounts and patterns usually indicate additional human impacts on the basin [4,5] (NKounkou & Probst, 1987; Walling, 2009).

Wetlands are complex ecosystems that act as an interface between terrestrial and aquatic ecosystems. Wetlands receive water and sediments from external sources [6] (Mitsch & Gooselink, 2015). The hydrodynamic characteristics of wetland ecosystems, including the water inputs, outputs, water flow, and seasonality, can influence organic deposition and material fluxes [7] (Gosselink & Turner, 1978). [8] Novitzki (1979) charac-terized wetlands with regard to water source and landform as surface or groundwater depressions or slopes, which are defined by whether or not they receive a greater amount of water from precipitation and overland flow or by their intersection of the water table. Therefore, the type of wetland will play a significant role in the sediment dynamics of a catchment, as one of the specific functions of wetlands is serving as a sediment trap [6,9,10] (Laraque et al., 2009; Mitsch & Gooselink, 2015; Mitsch et al., 2015).

The Congo River Basin (CRB) is the second-largest continuous tropical rainforest in the world and hosts the largest peatland complex found anywhere in Africa [11] (Dargie et al., 2017). Peatlands are the largest natural terrestrial carbon store, helping to minimise flood and drought risk, retaining sediment and preserving global biodiversity [12] (Harenda et al., 2018). The central depression in the heart of this basin, famously called the "Cuvette Centrale", encircles the peatland complex. There are important spatial variations in the tributaries that supply the Cuvette Centrale with water, sediments and nutrients. These tributaries are located both on the right bank and on the left bank tributaries of the main Congo River. These variations are due to the position of the basin straddling the equator, which results in different precipitation patterns on both sides of the hemisphere. A few authors have studied the material fluxes of the Congo basin [9,13–20] (Probst, 1983; Probst et al., 1992; Moukolo et al., 1993; Gaillardet et al., 1995; Dupré et al., 1996; Guyot et al., 2000; Seyler et al., 2006; Laraque et al., 2009; Mushi et al., 2019). These studies have quantified fluxes as part of a global study or estimated material fluxes at the outlets of major tributaries using hydro-sedimentary budgets based on in situ data, models or remote sensing methods. While they have laid appreciable foundations for further study, there remains a lot to be understood given the new information and dynamics regarding peatland stock and the changing precipitation patterns over some major tributaries occasioned by climate change [11,21,22] (Gumbricht et al., 2017; Dargie et al., 2017; Nguimalet & Orange, 2019).

The importance of the Cuvette Centrale cannot be over-emphasized. Given its role as a globally important asset in the fight against climate change, biodiversity loss, and social instability in the region. An analysis of the nature of wetland material, their particulate and dissolved fluxes, spatial and temporal variability could provide important indicators of climatic and environmental change and anthropogenic factors causing these changes. The sources of "Cuvette Centrale" sediment-laden waters, whether through smaller or larger streams, as well as the dynamics of the loads they carry, remains a grey area. In addition, erosional processes have not been clearly understood, while the spatio-temporal variations in sediment dynamics within the basin have more questions than answers [17] (Mushi et al., 2019). Several factors are responsible for the fluvial transport of sediments to the ocean. It is important to study them to better assess their effect on the environment at whatever scale [5,23] (Walling, 2009; Aagaard & Hughes, 2013). In addition to identifying potential sediment source areas, it is useful to estimate the volumes of sediment that are carried into the reach, as declining or increased sediment supply can influence wetland stability [24,25] (Oeurng et al., 2011; Guan et al., 2017).

This paper uses modelling tools to estimate sediment yields in the sub-basins and fluxes transported in the important tributaries of the Congo basin. While other studies have conducted independent assessments of the yields in specific sub-basins, we have considered the tributaries that feed into the Cuvette Centrale from both northern and southern tributaries taking note of the spatio-temporal variations in these tributaries. Specifically, this study aims to: (i) quantify the sediment fluxes in the CRB based on daily discharge and suspended sediment concentration; (ii) identify the important contributing tributaries as well as the critical source areas and (iii) estimate, for the first time, an upstream-downstream sediment balance within the Central Cuvette. This study is carried out for the 2000 to 2012 period, assuming that this period represents the hydro-sedimentary chronicle of the Congo basin [26–28] (Laraque et al., 2020; Moukandi N'kaya et al., 2020, 2021). It is also important to note that this study considers only the suspended sediments without the bedload component. The study also discusses the role played by the Cuvette Centrale in the hydro-sedimentary budget of the Congo basin.

2. Materials and Methods

2.1. Study Area

2.1.1. General Characteristics

The CRB is the second largest river basin and the second largest area of contiguous tropical rainforest in the world. In Africa, it drains ten countries [29] (BRLi, 2016), partly or wholly, located both on the northern and southern hemispheres (Figure 1). At the Brazzaville/Kinshasa-gauging station, its size corresponds to 3.7 million km^2 and its mean discharge measured from 1902 to 2011 is 41,100 m^3 s^{-1} [29] (BRLi, 2016). The Congo basin is an intra-continental sedimentary basin encircled by relatively gentle relief (except for a few volcanoes on its eastern border, see Figure 2c), drained by numerous tributaries of the main Congo River [30] (Kadima et al., 2011). Its internal broad bowl-shaped sedimentary depression or "sag basin", generally referred to as the Cuvette Centrale [30–33] (Daly et al., 1992; Laraque et al., 1998b; Kadima et al. 2011; Alsdorf et al., 2016), results from the geodynamic influence of tectonic events [30] (Kadima et al. 2011). Its size is 1.2 million km^2 between 3°S-1°N latitude and 18–22° E longitude [31] (Daly et al., 1992), 32% (360,000 km^2) of which corresponds to wetlands [34] (Bwangoy et al., 2010).

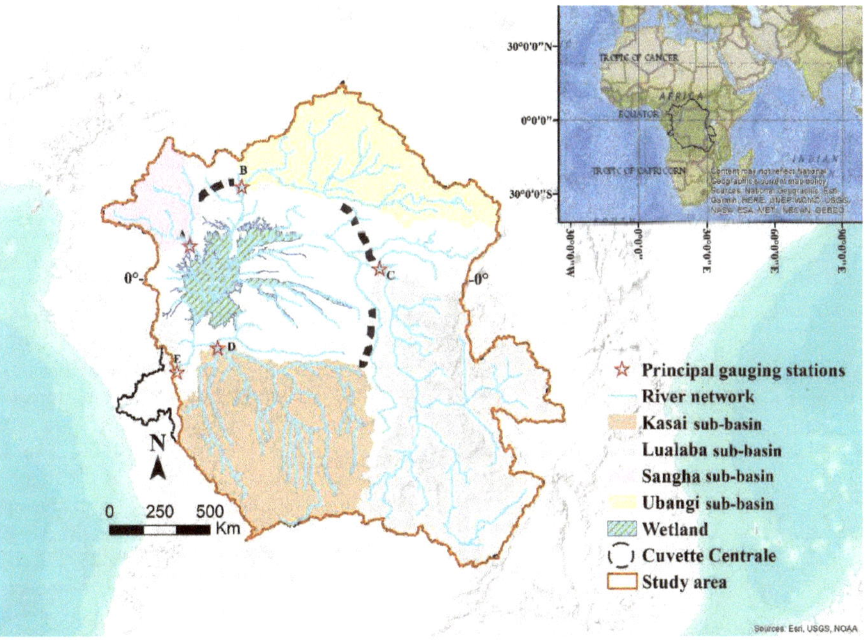

Figure 1. Map of the Congo basin showing its location in Africa, river network above the main gauging station of the Congo River at Brazzaville/Kinshasa (E), the principal gauging stations of Sangha at Ouesso (A), Ubangi at Bangui (B), Lualaba at Kisangani (C), and Kasai at Kutu Moke (D). The extent of the Cuvette Centrale (sensu lato) is also shown [26] (after Laraque et al., 2020), and the wetlands of the Cuvette Centrale were extracted from the global wetland database [35] (Lehner & Doll, 2004).

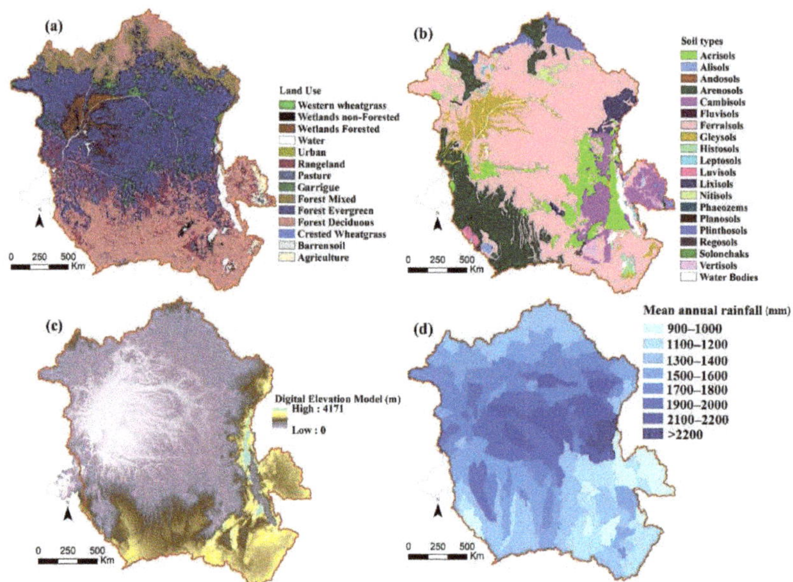

Figure 2. Maps of (**a**) land use (Global Land Cover 2000 database); (**b**) soils (Harmonized World Soil Database v 1.1); (**c**) Digital Elevation Model (Consortium for spatial information); and (**d**) simulated precipitation from the TRMM product (Multi-satellite precipitation analysis) used as inputs in the SWAT model. The map references and resolutions are listed in Table 1.

The main drainage systems of the basin are the Ubangi and Sangha catchments in the north and the Lualaba and Kasai catchments in the south, which are separated by the equator (Figure 1). Their tributaries converge and join the Congo main stem in the middle of the basin. Precipitation and hydrological regimes over the basin are bimodal and typical of equatorial regions, with the main flood from November to January and a lesser flood from April to June [28] (Moukandi et al., 2021). Annual average rainfall in the Cuvette Central are over 2000 mm yr^{-1} and decrease to as low as 1000 mm yr^{-1} in the northern and southern portions of the basin ([33] Alsdorf et al., 2016, see Figure 2d). The Cuvette Centrale is estimated to accumulate up to 9 km of sediment over much of geological time [36] (Crosby et al., 2010). The lithology of the basin comprises chiefly of alluvium, metamorphic and clastic sediments with most of the groundwater stored in permeable unconsolidated alluvium that could reach 400 m deep [37,38] (UNESCO, 2004; Schlueter, 2006). Dominant soil classes in the basin are ferralsols, arenosols, and nitisols (Figure 2b). Sands and clays dominate the basin soils with clay proportions dominating the Cuvette Centrale and the Sangha basin while sands dominate in the southern parts of the basin [17,39] (Mushie et al., 2019; Kabuya et al., 2020). The land cover of the basin consists of evergreen forests in the central basin with other uses such as agriculture, pasture and water bodies (Figure 2a).

2.1.2. Hydro-Sedimentary Measurements in the Congo River Basin

The Congo River Basin (CRB) is among the least studied river basins in the world [26,33] (Alsdorf et al., 2016; Laraque et al., 2020). Hydro-pluviometric networks have existed in the basin since colonial times. Some of the first analyses of material, both solid and dissolved, were carried out at the Brazzaville/Kinshasa gauge station in the early part of the 20th century. Discharge measurements have existed mainly at the Brazzaville/Kinshasa station, which controls 98% of the CRB. Monthly measurements of selected biogeochemical parameters have been carried out during the 1987–1994 period and continuously since 2005 until date. [26] Laraque et al. (2020) gives a chronological summary of the measurements of concentrations of Total Suspended Sediments (TSS), Total Dissolved Solids (TDS), and Dissolved Organic Carbon (DOC) performed since 1926 at the Brazzaville/Kinshasa gauging station.

The compilation showed that TSS concentrations remained relatively high during the first half of the century and through the wet decades of the 1960s, ranging from 12 mg L^{-1} to 58 mg L^{-1} from 1926 to 1986. [40] Olivry et al. (1988) attributed the dispersed results of earlier studies to the imprecision of measurements. [19] Probst et al. (1992) used stochastic and deterministic methods to estimate 41.61×10^6 t yr^{-1} and 3.41×10^6 t yr^{-1} of dissolved material exported by the Congo and Ubangi rivers 40 km upstream of Brazzaville and at Bangui respectively. They noted that the monthly concentrations of dissolved elements vary inversely with discharge. [13] Moukolo et al. (1993), on the other hand, derived values of 61.1×10^6 t yr^{-1} for the basin outlet from 1987 to 1991, using another method of calculation. For the next few years between 1987 and 1996, monthly sampling and analyses were performed and velocity weighted concentrations of TSS fluxes were computed [41,42] (Olivry et al., 1995; Coynel et al., 2005). The range of TSS fluxes was from 8.3 to 11 t km^{-2} yr^{-1} further highlighting the limited interannual variability of material fluxes in the CRB.

Laraque et al. [9,43] (2009, 2013a) gives a concise review of material export rates at the sub-catchment and whole catchment level respectively. The lack of data for some drainage units was overcome using the principle of similarity-a method that relies on a sound knowledge of the physiographic functioning of the basin. Mean TSS concentrations were also within the range of previous assessments at 25.0 mg L^{-1}. Similarly, [43] Laraque et al. (2013a) obtained, using linear interpolation methods for missing data, average TSS concentrations of 25.4 mg L^{-1} at the basin outlet. [27] Moukandi N'kaya et al. (2020) made a comparison of the variability between two measurement periods: the Programme d'étude de l'Environnement et de la Géosphère Intertropicale-Operation Grands Bassins Fluviaux (PEGI-GBF), and the Observation Service For The Geodynamical, Hydrological and Biogeochemical Control of Erosion/Alteration and Materials Transport in the Amazon, Orinoco, and Congo Basins (SO-HYBAM). They noted a 7.5% increase of TSS (from 30.2 to 33.6 $\times$ 10^6 t yr^{-1}), a 28% increase of DOC, and a 15% decrease in TDS concentration in the current period. They attributed these variations to the increase in water flows since the PEGI period. While these studies have shown the low variability of the material fluxes of the CRB, largely due to its stable and regular hydrological regime, they noticed an increase in the year-to-year variability of TSS concentration, from 25.3 ± 4.6 mg L^{-1} in 1987–1993 to 27.2 ± 7.9 mg L^{-1} in 2006–2012. However, there is the need for a more regular monitoring of the river networks with emphasis on the spatial distribution. For now, the only reliable gauge network is at the basin outlet with other sources of data, mainly independent studies conducted by workers for short periods. The obvious impact on this work will be the uncertainties that will arise from this lack of spatial and temporal representation.

2.2. Data and Methodology

The TRMM rainfall product (TMPA) 3B42 V.7 was used to force the SWAT hydrological model. The TRMM product has a spatial resolution of 0.25° and has been validated over the Congo basin by Munzimi et al. (2015) using independent datasets of historical isohyets. Water discharge and suspended sediment measurements from the five principal drainage basins of the Congo basin were obtained from the SO/HYBAM Observatory and its collaborators; notably the CICOS (Commission Internationale du bassin Congo-Oubangui-Sangha), the GIE-SCEVN (Groupement d'Intérêt Economique—Service Commun d'Entretien des Voies Navigables), the RVF (Régie des Voies Fluviales), the PEGI/GBF project, and the Marien N'gouabi University. The 90m resolution SRTM digital elevation model from the consortium from spatial information was used. This resolution was considered considering the size of the basin and the uncertainties related to DEMs [44] (Wechsler et al., 2007). Meteorological data from 1979 to 2014 was provided by the Climate Forecast System Reanalysis (CFSR) and is available at https://swat.tamu.edu/ (accessed on 6 June 2018). Other spatial data include the 1 km resolution Harmonized World Soil Database [45] (HWSD), while the land cover and land use information was derived from the Global Landcover database 2000. Further details of these inputs and other supporting data used are shown in Table 1.

Table 1. Input data used to run the SWAT model.

Data Type	Period	Resolution	Source
Land elevation		90 m	Digital Elevation Model. Consortium for spatial information (https://cgiarcsi.community/data/srtm-90m-digital-elevation-database) (accesed on the 6 June 2018)
Soil		1 km	Harmonized World Soil Database v 1.1 (http://webarchive.iiasa.ac.at/Research/LUC/External-World-soil-database/HTML/index.html?sb=1) (accesed on the 6 June 2018)
Land use		1 km	Global Land Cover 2000 database (http://forobs.jrc.ec.europa.eu/products/glc2000/products.php) (accesed on the 6 June 2018)
Rainfall	1998–2015	0.25°	TRMM (TMPA) 3B42 V.7 Daily product. Multi-satellite precipitation analysis (https://pmm.nasa.gov/data-access/downloads/trmm#) (accesed on the 6 June 2018), [46] Huffman et al. (2007)
Meteorological data	1979–2014	~38 km	Climate Forecast System Reanalysis (CFSR) Model (http://rda.ucar.edu/pub/cfsr.html&http://globalweather.tamu.edu/) (accesed on the 6 June 2018)
River discharge	2000–2012	Daily	SO-HYBAM (http://www.so-hybam.org/) (accesed on the 6 June 2018); [30] BRLi (2016)
Suspended sediments	2006–2017	Monthly	SO-HYBAM (http://www.so-hybam.org/) (accesed on the 6 June 2018)
Supporting data			
Water productivity	2009–2018	250 m	FAO (https://wapor.apps.fao.org/catalog/WAPOR_2/1) (accesed on the 6 June 2018)
Wetland extent	1992–2000	30 × 30 s	Global Wetlands database, [36] Lehner and Döll (2004) (https://www.worldwildlife.org/publications) (accesed on the 6 June 2018)
Geology	2012	0.5°	Global lithological database [47] (Hartmann and Moosdorf, 2012).

2.3. Modeling Approach

SWAT Model Overview

The Soil and Water Assessment Tool (SWAT) is a semi-distributed, physically-based model capable of integrating various spatial environmental data, including climate, soil, land cover and topographic features. SWAT simulates various hydrological processes e.g., surface runoff, infiltration, evapotranspiration (ET), lateral flow, percolation to shallow and deep aquifers and channel routing [48] (Arnold et al., 2012). SWAT simulates the hydrology of a catchment in two ways: (i) the land phase of the hydrologic cycle and (ii) the channel or routing phase of the hydrological cycle. The first phase controls the amount of water, sediment, nutrients and pesticides loadings to the main channel in each subbasin. The second phase is related to the movement of water, sediments, nutrient and pesticide through the channel network of the watershed to the outlet [49] (Neitsch et al., 2011). More details on the hydrological setup of SWAT for the Congo basin can be found in [50] Datok et al. (2021) and further details of the hydrological processes and equations used for hydrological simulation can be found in the SWAT documentation (https://swat.tamu.edu/; accessed on 6 June 2018).

2.4. SWAT Hydrological Model Setup

The catchment was delineated based on the dominant land use, soil and slope classes apportioning one sub-basin per Hydrologic Response Unit (HRU). This delineation performed within ESRI ArcGIS 10.4 using the ArcSWAT interface (www.esri.com; accessed on 6 June 2018) resulted in 272 subbasins and HRUs with 20 land uses and 14 soil classes (Figure 2a,b). Reservoirs were integrated into the upper part of the Lualaba subbasin to take into account the series of lakes and swamps that act as storages and affect downstream flows. The reservoirs were placed in the outlet of the sub-basins that receive flow from Lakes Tanganyika, Upemba, and Mweru. Reservoir parameters were set according to [50] Datok et al. (2021). Surface runoff was simulated using a modification of the soil conservation service Curve Number (CN) method. The runoff from each sub-basin was routed through the river network to the main basin outlet using the variable storage method. We modelled Evapotranspiration using the Penman–Monteith method. The model was calibrated with measured data on a monthly time scale from 1998 to 2012; comprising calibration (2000–2006), validation (2006–2012), and a two-year warm-up period (1998–2000), to enable the model to achieve a quasi-steady state condition. The criteria of efficiency (NSE) and Kling–Gupta efficiency [51] (Gupta et al., 2009) in 60% of the calibrated and validated stations were above 0.6 and 0.7, respectively, while the percent bias (PBIAS) values ranged from −8 to 8 in all the stations. This result further gave confidence that the hydrology was well simulated, therefore guaranteeing a solid base for estimating soil erosion and sediment dynamics. Further details on the hydrological simulations performed in this case study can be found in [50] Datok et al. (2021).

2.5. SWAT Sediment Model

SWAT uses a Modified Universal Soil Loss Equation (MUSLE) developed by [52] Williams (1975) to simulate sediment yield:

$$Sed = 11.8 \times (Q_{surf} \times q_{peak} \times A_{hru})^{0.56} \times (C_{USLE} \times P_{USLE} \times K_{USLE} \times LS_{USLE} \times F_{CRFG}) \tag{1}$$

where, Sed is the sediment yield on a given day (metric tons day^{-1}); Q_{surf} = Surface runoff volume (mm day^{-1}); q_{peak} = runoff peak discharge (m^3 s^{-1}); A_{hru} = HRU area (ha); C, P, K, LS and F, are non-dimensional factors associated with the crop cover, management practice, soil erodibility, slope and stoniness [53] (Wischmeier and Smith, 1978). The stoniness refers to the fraction of rock fragments at the surface and must be distinguished from rock fragments below the surface. Surface rock fragments affect the C-cover factor, while below surface rock fragments affect the soil erodibility. The MUSLE replaced the rainfall factor (R) of the original USLE [54] (Wischmeier and Smith 1965) with instantaneous peak flows and total runoff to predict soil erosion. Two processes control the sediment transport in the channel network: deposition or erosion. The channel dimensions can either be kept constant so that SWAT can compute deposition/erosion. Otherwise, channel degradation can be activated which would allow channel dimensions to change and be constantly updated as a result of downcutting and widening [49] (Neitsch et al., 2011). In this study, we selected the former option, as the latter has not been proven yet.

Just like the hydrological phase, the sediment phase also consists of the landscape and channel component. SWAT routes different particle sizes through grassed waterways, wetlands/ponds and vegetative filter strips in the land phase. Therefore, the sediment yield reaching the stream channel is the sum of the total sediment yield calculated by MUSLE minus the lag and the impediments listed before [49] (Neitsch et al., 2011). To calculate sediment balance, Sediment from upland sub-basins are routed and added to downstream reaches. This depends on the sediment transport capacity of the reach, which in turn depends on the stream power at the reach. A modified Bagnold's equation is employed to calculate the maximum concentration of sediments that can be transported in the reach:

$$Conc_{max} = C_{sp} \left(\frac{prf \times q_{ch}}{A_{ch}} \right)^{spexp} \tag{2}$$

where $\mathrm{Conc_{max}}$ = the maximum daily concentration of sediments (t m^{-3}) that can be transported in the reach; *prf* = the sediment peak rate adjustment factor; q_{ch} =average rate of peak flow (m^3 s^{-1}); A_{ch} = channel cross-sectional area (m^2). The coefficients C_{sp} and spexp regulate the linear and exponential relationship between the stream power and the peak velocity and are customized by the user during model calibration. Sediments are deposited when sediment concentration exceeds $\mathrm{Conc_{max}}$ (aggradation). Conversely, when sediment concentrations are below the stream transport capacity, nothing is deposited and bed sediment could be eroded if channel erodibility is activated. In this study, channel erodibility was applied sparingly during calibration in selected subbasins. More details of sediment transport equations and processes simulated in SWAT can be found in the SWAT theoretical documentation [49] (Neitsch et al., 2011).

2.6. SWAT Sediment Setup

SWAT2012 and ArcGIS 10.4 were used in this study. As stated earlier, the whole basin was divided into 272 subbasins with 14 land-use classes (Figure 2a) and 20 soil classes (Figure 2b). Only one HRU was defined for each subbasin, based on the combination of the dominant land use and soil type. This was done considering the size of the basin and the computational requirements needed. Furthermore, the MUSLE must be applied to a hydrologically isolated unit for a correct assessment of the water balance [55] (Vigiak et al., 2015). The final range of the HRU area was 0.085–61,775 km^2 with a median value of 9726 km^2. We considered this delineation fit for our purpose, with consideration of our inputs, computational resources and modelling priorities to avoid the risk of misleading results [56] (Jha et al., 2004).

The USLE cover and management factor (C_{USLE}) values were converted from the 14 land uses (Figure 2a) by SWAT based on the amount of soil cover specified for a particular land use, and were extracted from the soil database during model set up. The agricultural area represents a small part of the land use in the basin (<2%) (see Figure 2a), therefore the C_{USLE} and support practice (P_{USLE}) factors were not calibrated and set at default values over this land use. The topographic factor LS_{USLE} is automatically derived from the DEM while the coarse fragment factor (F_{CRFG}) is calculated from the percent of rock in the first soil layer. When there is no rock in the first layer, the F_{CRFG} value is 1.

The soil erodibility factor (K_{USLE}) was carefully calculated over each HRU. This was important in order not to expect too much from default model parameters, and in data scarce basins such as the CRB; a hydrological model should be setup with as much information as available. The default model parameters are based on the Harmonized World Soil Database, and SWAT commonly uses the most dominant soil series in that soil association. The information contained in the database is variable [45] (HWSD), and thus may or may not be representative of the soils in the locality.

The K_{USLE} erodibility factor K was estimated using the freely available software Kuery 1.4 [57,58] (Borselli et al., 2009, 2012). It is based on the equations developed by [59,60] Torri et al. (1997, 2002) from observations of a global dataset of 240 soils where erodibility is expressed as:

$$K = 0.0293\left(0.65 - D_g + 0.24D_g{}^2\right)e^{\{-0.021(\frac{OM}{C})-0.00037\,(\frac{OM}{C})^2-4.02C+1.72C^2\}} \tag{3}$$

where C is the fraction of total clay content, OM is organic matter as a percentage, and D_g is the logarithm of the geometric mean D of the particle size distribution. In cases such as our own, where data of textural components of soil exist only as a percentage; sand (S), loam (L) and clay(C), D_g is calculated according to [61] Shirazi et al. (1988) as:

$$D_g = \frac{-3.5C - 2.0L - 0.5S}{100} \tag{4}$$

As compared to a homogeneous soil, K is affected by the percentage of rock fragment (R_{fg}) in the soil, which is needed as input in Kuery 1.4, as well as the climate classification of the soils [62] (Salvador Sanchis et al., 2008). In addition, the logarithm of the geometric standard deviation of the geometric mean of the particle size distribution (Sg) [61] (Shirazi et al., 1988) is needed:

$$S_g = \sqrt{\frac{C\left(-3.5 - D_g\right)^2 + L\left(-2.0 - D_g\right)^2 + S\left(-0.5 - D_g\right)^2}{100}} \tag{5}$$

The technical report of the soil and terrain database of the Democratic Republic of the Congo [63] (Engelen & Verdoodt, 2006), the Soil and Terrain Database for Central Africa (SOTERCAF) and the Harmonized World Soil Database [45] (HWSD) were consulted for the basic inputs D_g, S_g, OM, and percentage rock content (R_{fg}). The means of the resulting probability distribution was implemented and spatialized in the SWAT model based on the dominant soil types in the basin (Figure 2b). Table 2 shows the textural components in the percentage of ferralsols used as inputs to the Kuery 1.4 model. Ferralsols are the dominant soil class in the CRB covering 58% of the watershed area as delineated by the SWAT model. Textural components of the ferralsols were obtained based on the soil characteristics of eight locations covered by these soils. Other dominant soil classes with their textural components are shown in Table 3. These eight soil classes make up 40% of the watershed area with watershed size ranging from 0.65% to 17.41%. The textural component percentages of these eight soil classes were obtained from characteristics of these soils and generalized across the basin (Table 3). The climate zone for our study area corresponds to the tropical classification of the Köppen–Geiger climate map of [64] Peel et al. (2007) and identified as such in our study. The generated K values frequency distribution allowed us to choose a more representative value for a given soil based on its characteristics given in our soil databases. These characteristics linked mainly to soil structural stability are derived from the description of soil profiles in the databases and literature.

Table 2. Percentage values and statistics of textural components of ferralsols, which make up 58% of the basin soils, used in the Kuery 1.4 model.

	Sand%	Loam%	Clay%	Organic Matter%
MIN	17.7	17.9	0.8	0.52
MAX	49.9	69.4	36.6	11.1
MEAN	31.20	42.79	22.79	2.35
Std ± (standard deviation)	11.36	18.48	12.07	3.64

Table 3. Values of textural components of other dominant soils in the basin used in the Kuery 1.4 model.

SOIL	Percentage of Watershed Area	Sand%	Loam%	Clay%	Organic Matter%
Acrisols	7.0	60	20	20	0.4
Arenosols	17.41	70	20	10	0.7
Cambisols	6.98	47.9	31.5	20.6	0.46
Gleysols	3.74	10	20	70	7.2
Lixisols	1.21	8.2	44.5	47.3	1.2
Luvisols	0.80	73.8	20.7	5.5	0.18
Nitisols	0.65	10	30	60	2.17
Plinthosols	2.65	52	28	20	0.25

2.7. Sediment Balance

We replicate the approach used by [50] Datok et al. (2021) in their hydrological study to estimate the sediment balance in the Cuvette Centrale. 11 points were selected in the model; eight upstream sub-catchments of River stations with their outlets feeding into the wetlands of the Cuvette Centrale (stations 1 to 8, Figure 3). They were completed by three additional stations: one at the outlet of the Kasai subbasin (st. 10), one at the gauging station of Brazzaville/Kinshasa (st. 11), and one intermediate station in the Congo River downstream of the wetlands (st. 9). The total input of sediment for the Centrale Cuvette is given by $I = \sum_{i=1}^{8} S_i$, the output is considered as the flux at Brazzaville/Kinshasa minus the contribution of Kasai $O = S_{11} - S_{10}$, and the sediment balance of the Cuvette Centrale is:

$$\Delta S = I - O \tag{6}$$

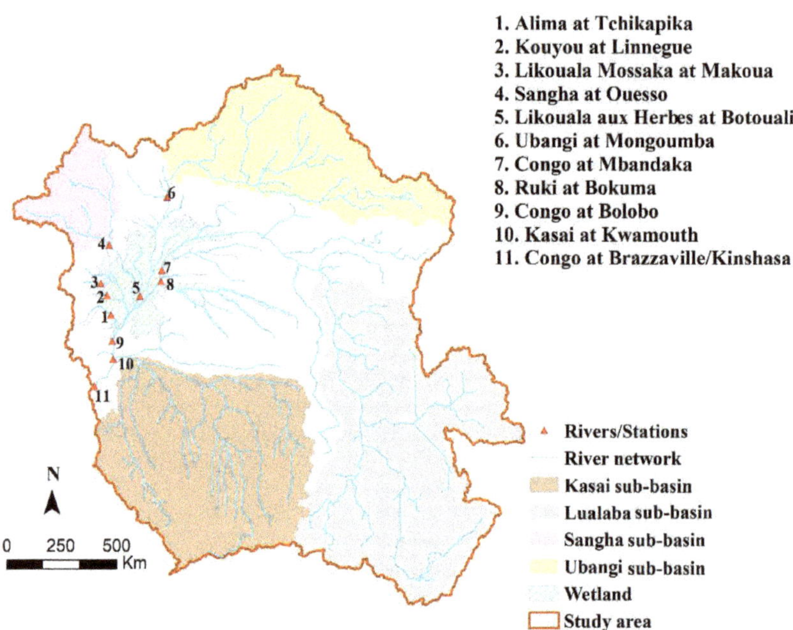

Figure 3. The Congo River Basin showing points used in the calculation of the sediment balance numbered 1 to 11.

2.8. Calibration and Validation of the Model

The hydrological model was validated daily. The model was then calibrated on a daily time-step with suspended sediment (SS) concentration time series (mg L^{-1}). This time series comprises a one to three times monthly sampling obtained from the HYBAM observatory located at the Brazzaville/Kinshasa gauging station. The period of the calibration was 2006 to 2009, while validation was performed from 2010 to 2012. The other only source of SS data that was reliable for use was that measured on the Ubangi River [65,66] (Bouillon et al., 2012, 2014). We used their dataset to calibrate and validate the model at the upstream station on the Ubangi River at Bangui: from March 2010 to March 2011 for calibration, and from April 2011 to March 2012 for validation. The Brazzaville/Kinshasa records used were once-monthly readings from 2006 to 2012 while the measurements on the Ubangi River were based on the 52-day sampling program of [65,66] Bouillon et al. (2012, 2014). Spatialization of the calibration was highly limited due to the lack of gauging stations; therefore, we relied on literature data to calibrate broadly. Manual calibration with the aid of expert knowledge was also used for calibration. Since we lacked critical data for calibration, we first ran our model as simple as possible. We assumed zero net erosion or deposition in the channel by setting the exponential and linear parameters to eliminate deposition and the channel routing parameters to eliminate erosion. We then adjusted the support practice factor that controls the upland sediment yields and calibrated parameters controlling channel deposition and erosion in an upstream-downstream manner. Details are given in the next section.

Due to the paucity of SS data in this region, the simulated SS concentrations in the period 2000–2012 were validated using historical records measured at the Brazzaville/Kinshasa gauging station from the PEGI/GBF program (1987–1994). The justification for this method is based on the works of [67,68] Laraque et al. (2001, 2013b) and [28] Moukandi et al. (2021), who using statistical tests highlighted the low variability of hydrological regimes during different homogenous periods in the CRB. We selected stations that had at least five months of readings and compared the means noting the assumptions made in comparing the model performance objectively due to the lack of a common frequency of measurement.

2.9. Statistical Analysis

Literature has described the CRB as flat [31] (Daly et al., 1992), with the absence of mountain ranges and thick vegetation cover, resulting in low erosion rates [26] (Laraque et al., 2020). Therefore, a Principal Component Analysis was carried out to identify the most influential factors affecting sediment yield and erosion in the CRB. [69] Vanmaercke et al. (2014) analysed sediment yield in Africa and showed that seismicity, topography, vegetation cover and runoff are the main factors that control sediment yield. Therefore, based on the influence of the parameters analysed in the hydrological simulation in [50] Datok et al. (2021), as well as the well-documented influence of slope and soil on erosion in the CRB [17,39] (Mushi et al., 2019; Kabuya et al., 2020), five variables were incorporated in the PCA. They are the slope in the HRUs (HRU_slp), precipitation (PCP), soil erodibility, (USLE_k), channel vegetation cover (CH_COV2), and surface runoff (SUR_Q). A Pearson correlation was also made between the variables and the most important principle components to measure the strength of their association. These analyses were performed in Rstudio with R version 4.0.

Quantitative means of assessment were also used to evaluate model performance. They include the Nash Sutcliffe efficiency (NSE), the coefficient of determination (R^2) and percentage bias (PBIAS) [70] (Moriasi et al., 2015).

3. Results

3.1. Hydrological Responses in the Main Tributaries

Generally, the model exhibited an acceptable correspondence between the observed and modelled flows, showing its ability to capture the general shape of the hydrograph and magnitude of both high and low flows satisfactorily, at the outlet of the basin as well as in its sub-basins [50] (Datok et al., 2021). The validation of the daily simulations highlighted that SWAT can be calibrated at a monthly time step and validated at a daily time step. The differences in the evaluation criteria show the effect of the regionalization in the calibration as well as the level of complexity of each sub-basin. The lower performance of the daily model compared to the monthly simulation can be linked to the inadequate spatial representation of rainfall as well as errors in other inputs and measurements [71] (Adla et al., 2019). In addition, there are obvious differences between the observation and simulation datasets in both time steps. Much of the discussion about the hydrological responses in the various sub-basins have been documented in [50] Datok et al. (2021). However, to have an overview, we summarize the main points that are essential for the application to sediment modelling.

The Northern sub-basin of the Ubangi, which controls about 16% of the basin, gave good to satisfactory outputs for the evaluation criteria used since it had the most reliable observed data amongst all sub-basins. Therefore, most of the processes operating in the basin were well captured.

The Sangha sub-basin, which is situated west of the Ubangi sub-basin (Figure 1), was also calibrated on a monthly time scale. The model was not able to capture the minor flood phases in July and August, which were amplified on a daily scale. This was put down to the effect of wetland processes and the inadequacies of the model structure to deal with this. Nevertheless, it reproduced satisfactorily the timing of the hydrograph as well as the water balance components.

In the southwestern Kasai sub-basin, performance ratings ranged from satisfactory to good, with the hydrograph dynamics well captured. The percent bias (PBIAS) in this sub-basin was low, which could be deceptive as the model underpredicted as much as it over-predicted. However, other evaluation criteria (the NSE, R^2), gave a good account of the results here, and the discharge predicted was within three per cent of the observed values.

The challenge was in attenuating the flows from the wetland regions of the basin, specifically in the upper parts of the Lualaba and the central basin. Reservoirs were added to the Lualaba basin to help in trapping the water generated by the large water bodies there. However, we were able to capture satisfactorily the magnitude of flows within one per cent of the observed value.

At the basin outlet, the effect of the convergence of flows from all the contributing tributaries was evident. We were able to reproduce the magnitude of the flows to within 9% of the observed flows. The objective of the study is not to achieve a perfect calibration of the discharge at the outlet, as the challenges have been well documented [72–75] (Chishugi & Alemaw, 2009; Tshimanga et al., 2012, 2014; Munzimi et al., 2019). Rather, the objective is to find a balance between the flows received by the central basin before being impacted by the "Cuvette Centrale" and the flows measured at the basin outlet. Moreover, to achieve this aim using a mass balance approach, observed measurements are available at the outlet for comparison. An analysis of the parameters showed that groundwater parameters were influential all across the basin, with subsurface processes dominating the Sangha and the central basin. The central basin also had the least response to recharge, while the Kasai and Ubangi basins are drawing up more water from the shallow aquifer to the overlying unsaturated zones through deep-rooted vegetation. Some optimal parameters like the canopy interception compared well with physical observations, being higher in areas with more vegetation and lower in less vegetated areas.

In estimating the water balance of the Cuvette Centrale, [50] Datok et al. (2021) found that the main source of Cuvette Centrale waters was from precipitation (31%) with upland runoff contributing approximately equal amounts (33%) for the 2000 to 2012 simulation period. The Ubangi is the largest tributary contributing 16% while the other left bank tributaries contributed a combined 11%. The seasonality of the flows was well captured, with the main flood in December and two smaller floods in March and May. The balance within the Cuvette Centrale is in surplus only during peak floods in October and April and in deficit at all other periods. The regulatory role of the Cuvette Centrale is demonstrated by its ability to augment flows through groundwater and flooded areas to the River. More details of the hydrological model simulation outputs and results can be found in [50] Datok et al. (2021).

3.2. The Calibration Dataset for Sediment Transport and the Model's Calibration Performance
3.2.1. Data Analysis

Figure 4a shows the relationship between the hydrograph and the suspended sediment concentration on the Ubangi River for a one-year period from March 2010 to March 2011. The hydrological regime of the Ubangi is unimodal, with the main flood period from September to November. There is a noticeable lag in the suspended sediments as they follow the discharge, a form associated with a higher concentration-discharge ratio at the rising limb than the falling limb [76] (Megounif et al., 2013), creating a clockwise hysteresis (Figure 4b), described by [77] Laraque and Olivry (1996). The sediment decline at the peak flood is usually attributed to diminishing sediment available in the stream channels [78] (Wang et al., 1998). It must also be noticed that the variability of monthly-suspended sediment concentration (SSC) values is high in the Ubangui basin (from 2.4 to 33.2 mg L^{-1} in 2010–2011).

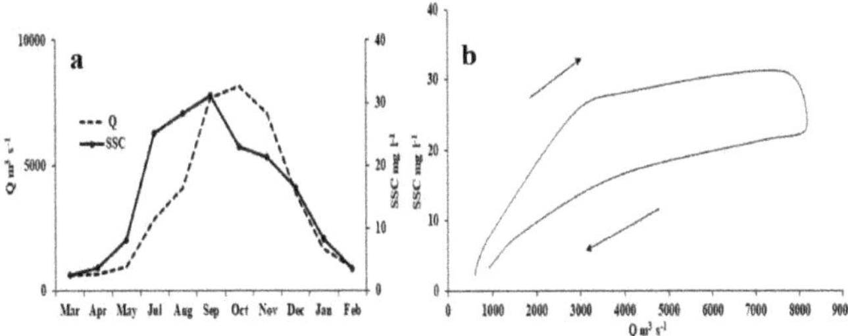

Figure 4. (**a**) The hydrograph and sediment concentration at the Sampling site on the Ubangi River from March 2010 to March 2011 [65] (Bouillon et al., 2012), and (**b**) the resulting hysteresis pattern.

An analysis of the interannual monthly sediment dynamics from 2006 to 2012 at the Brazzaville/Kinshasa gauging station shows the highest concentrations of suspended sediment in March to May at low water while the lowest concentrations are in December/January during the highest peak floods (Figure 5a). [43] Laraque et al. (2013a) also calculated a higher mean monthly concentration of SS of 31.9 mg L^{-1} in March and a lower concentration of 19.1 mg L^{-1} in December for the 2006 to 2010 period. [77] Laraque and Olivry (1996) noted the absence of the erosion-transport-alluvial phases that are associated with the Bangui station. Instead, the Brazzaville/Kinshasa outlet is noted for a single transport phase. This is due to the low seasonal variations in its particulate concentrations and its flows due to the all-year-round contributions of different sub-basins of the watershed to its flow. The hysteresis in Figure 5b displays a figure-eight pattern that combines a clockwise and anti-clockwise stage. This is due to the influence of the two equatorial regimes consisting of the main flood peak and a lesser one. The clockwise stage involves an increase of concentrations from the erosion of initially deposited stream sediments and thereafter a dilution effect with increased discharge [78,79] (Laraque et al., 1995; Aich et al., 2014). The anticlockwise stage follows with decreased discharge and increased sediment concentration on the falling limb. This is due to more available sediment for transport [76] (Megnounif et al., 2013), and in this case, could be related to travel time and contribution of sediment from upstream sub-basins and/or the effect of upstream storages like the Cuvette Centrale [78,80] (Laraque et al., 1995; Malutta et al., 2020). [27] Moukandi N'kaya et al. (2020) analysed the concentration-discharge curves from 2006 to 2017 at the same station and noted a classic dilution pattern of concentration by the discharge with the dilution and concentration effect well marked at the peak and low floods respectively. This figure-eight pattern is common to watersheds like the CRB with at least two distinct flood peaks.

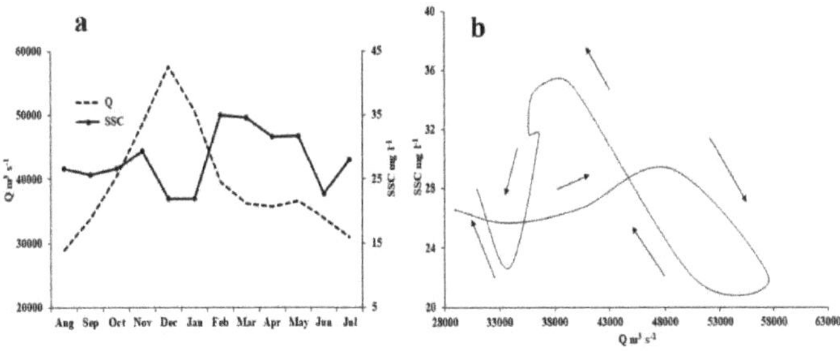

Figure 5. (**a**) The hydrograph and sediment concentration recorded at the Brazzaville/Kinshasa gauging station (2006–2012), and (**b**) the resulting hysteresis pattern.

3.2.2. Calibration Performance

Table 4 shows some parameters used to calibrate sediment dynamics in the SWAT model. The parameters that proved important in the calibration of the sediment yield and transport are related to the channel processes and processes in the Hydrological Response Units (HRUs). The simulation reveals that the exponential parameter for sediment routing in the channel was the main parameter among the general basin-wide parameters with influence. The values of channel erodibility (CH_COV1) and sediment transport coefficient (SPCON), set to their default values reveal the dominance of channel depositional processes within the basin. The calibration of these factors was done in an upstream-downstream manner. A filter was also implemented in sub-basin 168 corresponding to the confluence of the Kasai with the main river. This was important, as this is a site of deposition due to a change in velocity [17] (Mushi et al., 2019). The length of the filter simulated is 28 m and caused a reduction of about 4% of the total sediment yield at the basin outlet. It is also important to note that the filter strip reduces sediment but does not affect surface runoff [49] (Neitsch et al., 2011).

Table 4. Parameters used to calibrate the sediment model.

Parameters	Description of Parameter	Default	Applied Range
SPEXP	Exponential re-entrainment parameter for channel sediment routing	1	0.5
SPCON (C_{sp})	Linear re-entrainment parameter for channel sediment routing	0.0001	0.0001
PRF	peak rate adjustment factor for sediment routing in main channel	1	1
CH_COV1 (cm^3/N-s)	Channel erodibility factor	0	0–0.02
CH_COV2	Channel cover factor	0	0.65–1
CH_EQN	Simplified Bagnold Equation	0	1
Filter_W (m)	Edge of width filter strip (sub-basin 168)	0	28
LATSED (mg L^{-1})	Concentration of sediment in lateral and groundwater flow	0	50–60
K_{USLE} (t.ha.h./(ha.MJ.mm))	USLE soil erodibility factor	Relative	0–0.04
P_{USLE}	USLE support practice factor	1	1
C_{USLE}	USLE crop cover factor	Relative	0–0.2

Calibration was done at a daily time step at the Bangui station outlet using measured data from the study of [65,66] Bouillon et al. (2012, 2014) from March 2010 to March 2012 (Figure 6a) and at the Brazzaville/Kinshasa station using measured data from SO-HYBAM from January 2006 to December 2012 (Figure 6b). At the Bangui station, the model was able to reproduce the general shape of the hydrograph, capturing both periods of high exports and low exports, a reflection of the good responses already documented during the hydrological calibration and validation at this station [50] (Datok et al., 2021). Statistical metrics used to assess the models' performance are displayed in Table 5. The coefficient of regression of 0.69 showed how much of the variability of the suspended sediments are related to the discharge as the hydrological model has good skill scores. The Nash-Sutcliffe Coefficient of Efficiency (NSE) value of 0.35 is a satisfactory representation of the dynamics captured here when the temporal range of the observed data, as well as the sampling frequency, is considered. In Figure 6a, it can be seen that the simulated load does not follow the observed load during the rising phase of the hydrograph but is more correlated with the discharge at the falling limb. The abnormally high PBIAS value of 49.10 is because of this under-prediction of the rising limb where the magnitudes were not properly captured. Statistics were better in the validation period, with a satisfactory PBIAS value of 24.2%. It is also worth noting that in this sub-basin, the sediment discharge varies over more than two orders of magnitude (2.4–33.2 mg L^{-1} in TSS and 470–8540 m^3 s^{-1} in Q, resulting in a 101–24,508 t day^{-1} range for sediment discharge).

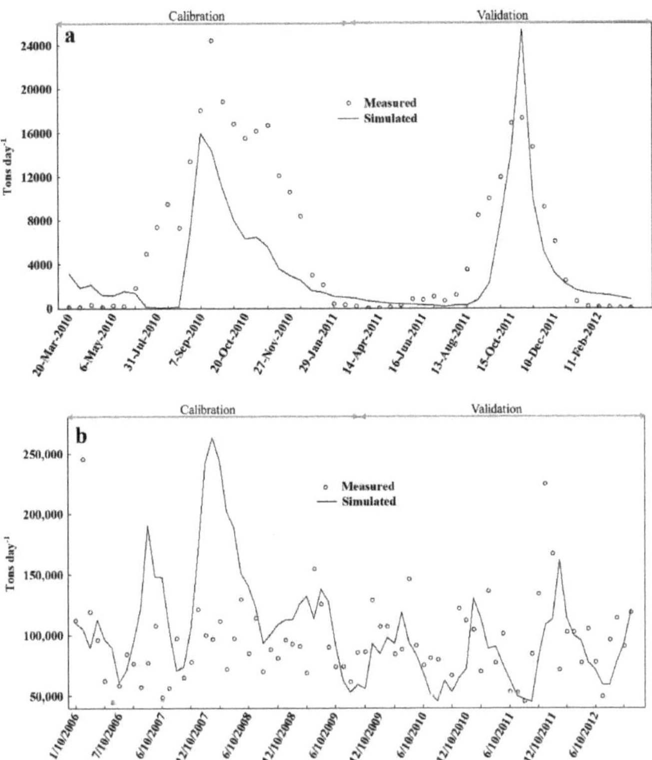

Figure 6. (**a**) Observed and simulated daily sediment fluxes at the northern sub-basin of the Ubangi River before the confluence with the main Congo River data from [65,66] Bouillon et al., 2012, 2014, and (**b**) Observed and simulated sediment fluxes at the basin outlet downstream of the Cuvette Centrale (data from SO-HYBAM).

Table 5. Performance indicators for the daily sediment flux (in t day^{-1}) calibration-validation periods of the model at the Bangui station and the Brazzaville/Kinshasa station.

Performance Metrics	Bangui		Brazzaville	
	Calibration (2010–2011)	Validation (2011–2012)	Calibration (2006–2009)	Validation (2010–2012)
NSE	0.35	0.65	−2.95	−0.16
R^2	0.69	0.71	0.02	0.13
PBIAS	49.10	24.20	−29.42	−9.53

Calibration and validation were also performed on a daily scale at the Brazzaville/Kinshasa outlet (Figure 6b). The Brazzaville/Kinshasa gauging station of the CRB receives flow from most of the tributaries that flow into the Cuvette Centrale as well as the Kasai, Plateaux Batekés Rivers and other tributaries downstream of the Cuvette Centrale. The calibration at this point highlights the necessity of a good hydrological calibration to achieve acceptable sediment calibration. Performance indicators at this outlet are −2.95, 0.02 and −29.42 for the NSE, R^2 and PBIAS respectively in the calibration period (2006–2009), while for the validation period (2010–2012), they are −0.16, 0.13, and 9.53 for the NSE, R^2 and PBIAS respectively (Table 5). The NSE is sensitive to extreme flows and reflects the under-prediction of peak flows as well as the influence of the Cuvette Centrale. The low values of regression coefficients also do not give a satisfactory indicator of the model performance at this outlet as the model did not accurately pick the timing of flows. This is in agreement with what has already been discussed about the failure of models to adequately slow the flow of water across the "Cuvette Centrale" [50,73] (Tshimanga et al., 2012, Datok et al., 2021). Lower statistical performance at Brazzaville Station (Figure 6b) than at Bangui (Figure 6a) also rely on much smaller ranges at Brazzaville (less than one order of magnitude) than at Bangui (more than two orders of magnitude in sediment load). The PBIAS of 9% in the validation period is adjudged satisfactory, however, showing that the model predicted well the average magnitudes although showing a bias towards underestimation of the fluxes.

From the analysis of our calibration dataset and the model calibration, it can be seen that SWAT sediment routing performs better on basins or sub-basins with low or moderate hysteresis (e.g., the Ubangi) than those with high hysteresis (e.g., at Brazzaville/Kinshasa). At Brazzaville, we observe a complex relationship between Q and SSC (figure-eight hysteresis), while the Bagnold formula considers a maximum concentration, increasing with increasing water discharge (or increasing discharge peaks). The difference may be attributed to processes of deposition/erosion over the land (especially in wetlands) or at the riverbed that are very difficult to calibrate correctly over a large basin with a very scarce dataset and natural complexities. To overcome these difficulties, there will be a need for additional measurements in the future at different stations of the basin to improve the model performance. In its state, the model was able to reproduce concentration values and sediment fluxes at a satisfactory level (with PBIAS < 50). It enables the first assessment of sediment balance in the Cuvette Centrale–which is the main target of the paper.

3.3. Additional Comparison of the Model Simulations with Historical Datasets

We compared our simulated fluxes with data from the PEGI-BGF program on the 1987–1994 period since there has not been any sustained sampling program over the entire basin. The justification of this comparison is based on the regularity of measured fluxes in the basin over the past century. For instance, [67] Laraque et al. (2001) highlighted that the hydro-climatology of the basin presented similar 10–12 year cycles in the last half-century. Ref. [81] Nguimalet & Orange (2013) also showed the homogeneity of rainfall-runoff ratios overtime at the Bangui station. Ref. [26] Laraque et al. (2020) showed that the magnitude of the discharge mainly controls suspended sediments due to their weak concentrations, also showing similar temporal variations as the hydrograph. Finally, [27] Moukandi N'kaya et al. (2020) compared the TSS concentrations for 1987–1993 with 2006–2017 values noting insignificant changes and concluded that their concentrations were similar. Simple statistical analysis of the variables in five locations are compared in Table 6. The PEGI observations of SSC and Q were sampled once every month, with some months not sampled at all in some stations, while we used the full time series for the SWAT simulations. More detailed statistical analysis could not be carried out because the frequency of measurement was not the same.

Average SSC values were very well estimated with the model at Bouatali (Likouala aux Herbes River), Ouesso (Sangha River) and Makoua (Likouala Mossaka River), with differences less than 7%, while it was over-estimated by 31.2% at BRZ/Kinshasa (Congo R.), and by 110% at Alima (Tchikapika) (Table 6). Of course, these quantities are not perfectly comparable since simulations ran over years while measurements were performed a limited number of times per station. However, this result confirms the ability of the model to assess a first sediment budget of the Cuvette Centrale.

Table 6. Descriptive statistics comparing suspended sediment concentration and discharge from the PEGI program (1987–1993) and SWAT simulation (2000–2012) from stations within the Cuvette Centrale.

					PEGI/GBF (1987–1993)				SWAT (2000–2012)				Difference between SWAT and PEGI Means (%)
River	Station	#	Parameters	n	Mean ± Std	Max	Min	Max/Min	Mean ± Std	Max	Min	Max/Min	
Alima	Tchikapika	1	SSC	28	7.3 ± 2.63	12.1	3.7	3.27	15.33 ± 1.06	17.06	10.91	1.56	110
			Q	33	553 ± 66.81	660	420	1.57	390 ± 103.91	401	275.8	2.85	29.5
Likouala Mossaka	Makoua	3	SSC	29	13.6 ± 5.77	30.9	3.9	2.95	14.55 ± 2.32	18.51	6.04	3.06	7.0
			Q	33	177 ± 79.73	307	62	3.15	91.55 ± 61.40	543	23.58	23	48.3
Sangha	Ouesso	4	SSC	7	18.5 ± 12.11	38.2	6.6	5.78	24.94 ± 6.57	58.83	15.19	3.87	34.8
			Q	14	1 155 ± 691.77	2920	514	5.68	1106 ± 374.18	4088	630	6.48	4.24
Likouala aux Herbes	Botouali	5	SSC	6	7.71 ± 2.59	11.00	4.60	2.39	7.45 ± 0.94	14.25	4.48	3.18	3.4
			Q	5	412.60 ± 185.18	546.00	134.00	4.07	327 ± 105.2	774	142	5.4	20.7
Congo	BRZ/KIN	11	SSC	99	23.99 ± 5.4	41.2	9.4	4.38	31.48 ± 7.44	78.11	18.66	4.18	31.2
			Q	99	38 515 ± 9642	45,000	37,100	1.21	36,156 ± 8221	64,690	18,430	3.51	6.1

n = number of observations; SSC = suspended sediment concentration (mg L^{-1}); Q = discharge (m^3 s^{-1}); Std = standard deviation; BRZ/KIN = Brazzaville/Kinshasa.

4. Discussion

4.1. Spatial Analysis of SWAT Outputs

4.1.1. Surface Runoff and Sediment Yield

We analysed the spatial variation of sediment yield across the basin (Figure 7). Generally, it can be observed that high surface runoff (Figure 7b) corresponds to high sediment yields (Figure 7a) across the subbasins. Surface runoff also followed the rainfall pattern of the basin with the central areas receiving some of the highest rainfall and above-average sediment yields. The sediment yields are generated from individual HRUs and from Figure 7a, it can be deduced that the Cuvette Centrale area with its various land uses including row cropped fields or pasture is characterized by local erosion. These sediments are generated and then accumulate here, aided by the low slopes and low velocities of water, which combine to restrict the transport of these sediments. The mapped sediment yield rates compare to the ones mapped by [9] Laraque et al., (2009). We estimated sediment yield rates of 4.01, 5.91, 7.88 and 8.68 t km^{-2} yr^{-1} respectively for the Ubangi sub-basin at Bangui, Sangha at Ouesso, Lualaba at Kisangani, and Kasai at Kutu-Moke respectively. The sources of these sediments are mainly from the higher slopes of the eastern highlands, the southern parts of the basin as well as sources from the Angolan highlands to the southwest. These loads are conveyed in the river from the upper course at an altitude of 1400–1500 m and they gather more mass as the river exits the numerous swamps and wetlands in the Lualaba basin. There is a noticeable reduction in river load as it makes its way through the anastomosing river channels of the middle Congo River. These channels are up to 16 km wide at some places [82] (Runge, 2007) and are characterized by sand and silt bars indicating deposition. Effective river width of around 5 km, measured 700 km upstream from Brazzaville/Kinshasa, reduce to less than 3 km at 1600 km upstream as measured by [83] Carr et al. (2019). At the equatorial regions, tributaries from the northern and southern hemispheres join the main river and increase the loads. Some of these loads are contributed by the Ubangi catchment due to the erosion of grazing land of the Sudano Sahelian environment [81] (Nguimalet & Orange, 2013). Water surface slopes are higher here with a reduced velocity. This river stretch is multi-channel with constrictions that reduce the widths considerably [83] (Carr et al., 2019). The Kasai River then joins the main stem at Kwamouth. The Kasai is noted to have a large suspended sediment load [9] (Laraque et al., 2009), contributing to the main river. This is consistent with measurements by [83] Carr et al. (2019) where they found water surface slopes to be very low in this area (2 cm km^{-1}). The loads are then transported and reduced through the Malebo pool with some loads produced at the urban settlement of Brazzaville/Kinshasa.

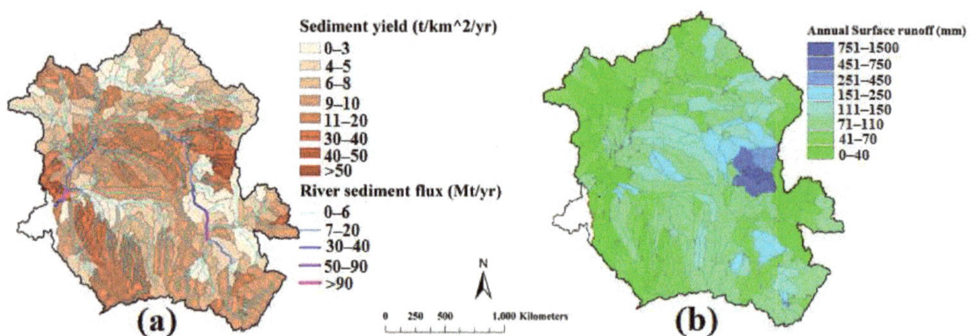

Figure 7. Mean annual variation of (**a**) sediment yield/flux and (**b**) surface runoff in the stream reaches of the CRB for the 2000–2012 period.

4.1.2. Land Use/Land Cover and Sediment Yield

We investigated the relationship between land cover, surface runoff and sediment yield in the CRB sub-basins over different land uses (Figure 8). The dominant land uses are forests, which occur in all the sub-basins while agriculture, pastoralism, and rangeland are the land uses with the most sediment yield.

The Ubangi sub-basin is dominated by forests of the mixed kind, comprising small trees and shrubs, which are interspersed with larger broadleaved trees at the northern limits of the basin. At the lower and wetter parts of the sub-basin are the boundaries of the evergreen forest. The highest yields of 4.5 t km^2 yr^{-1} was in the mixed forests located on higher slopes, while the lowest yield of 2.16 t km^{-2} yr^{-1} was in the evergreen forests. The Ubangi sub-basin presented the lowest specific sediment yield of the principal basins studied. However, in terms of load contribution, it ranks second of the sub-basins that contribute to the Cuvette Centrale after the Lualaba sub-basin. This is mainly because the discharge measured at the Bangui station is the largest of the Northern tributaries. Low sediment yields could also be attributed to the largely stable nature of the Ubangi tree-covered peneplains [17] (Mushi et al., 2019).

The Sangha basin drains the northwestern limits of the CRB; it has relatively high sediment yields compared to its size–5.91 t km^{-2} yr^{-1} (this study) and 8.52 t km^{-2} yr^{-1} [9] (Laraque et al., 2009). The highest yield of sediment is found in pastureland (12.22 t km^{-2} yr^{-1}, see Figure 8) while the evergreen forests produced the least sediment of 3.97 t km^{-2} yr^{-1}. Soils of the exposed pasturelands are more susceptible to erosion than those of the evergreen forests, which are protected by dense tree cover. This is a view held by [84] Ndomba et al. (2011), who also pointed out that agriculture and pastoralism are the main sources of sediment in catchments.

In the Lualaba sub-basin, agriculture is the largest land use by size with the highest sediment yields. Runoff is low though suggesting infiltration of water through the soil profile. Rangeland has a relatively high specific yield of 9.17 t km^{-2} yr^{-1} considering that this land-use ranks fourth in the area behind agriculture, evergreen, and deciduous forests. This may in part be due to overgrazing or other anthropogenic activities that expose the soil. Deciduous forests yielded 7.71 t km^{-2} yr^{-1} of sediment while the evergreen forests yielded 9.68 t km^{-2} yr^{-1} of sediment. The evergreen forests yield the least sediment in the other sub-basins studied. [85] Mayaux et al. (2003) described the evergreen swamp forests as "edaphic", alluding to the influence of the sediments on their sustenance. The trees are also mostly dense and multilayered, helping to slow the agents of erosion. The exception in the Lualaba may be due to the relatively higher slopes in some areas, which induce more yields. Non-forested wetlands yielded 1.25 t km^{-2} yr^{-1} of sediment while wheatgrass yielded 7.39 t km^{-2} yr^{-1} of sediment, noting that, while it is not as high as agricultural output, it shows the effect of different management practices on sediment yield.

The Kasai sub-basin is known to be heavily mineralized, due to the many mining activities taking place there. These anthropogenic activities are very impactful on the land. Therefore, it is not surprising that the highest sediment yields in the CRB are attributed to this sub-basin. We simulated specific sediment yields of 3.8 t km^{-2} yr^{-1}, 11 t km^{-2} yr^{-1}, and 12.7 t km^{-2} yr^{-1} for evergreen forests, deciduous forests, and rangeland, respectively (Figure 8). The sediment yield is more pronounced in the southwestern portion of this sub-basin probably due to anthropogenic activity mentioned earlier and/or less stable natural terrain [17] (Mushi et al., 2019).

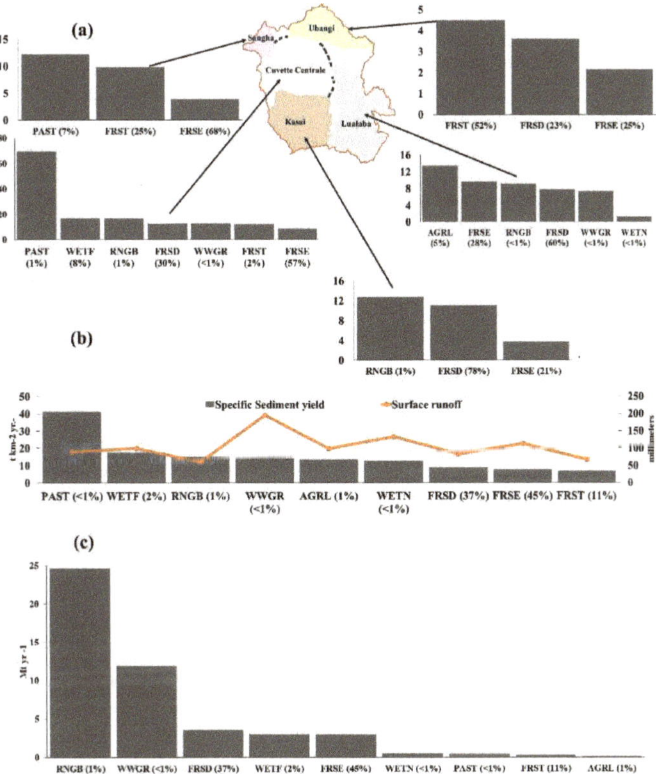

Figure 8. Mean interannual, (**a**) relationship between land cover and the upland sediment yield (t km^{-2} yr^{-1}) in the four subbasins of the CRB and the Cuvette Central area sensu lato (in brackets: percentage of each land cover in the sub-basin). (**b**) Variation of surface runoff with sediment yield (t km^{-2} yr^{-1}), by land use for the 2000–2012 simulation period over the whole basin. (**c**) Annual mean sediment loads produced by different land uses for the 2000–2012 simulation period over the whole basin. Note that the percentage of water bodies as the land cover is not included.

Land cover in the Cuvette is 57% Evergreen forest, 30% Deciduous forest, 8% Wetland forest, 2% Mixed forest with Pasture, Rangeland and Wheatgrass making up the balance. The soils are primarily composed of gleysols, which are affected by groundwater [86] (Ngongo et al., 2014). They are not well-drained and in fact, oversaturated causing relatively high surface runoff. We simulated the highest yields for the Pasture land, Deciduous forests, Wetland forests, Mixed forests, Evergreen forests, and Rangeland of 69, 14.3, 12.13, 12.09, 10.13, and 8.64 t km^{-2} yr^{-1} respectively. In the basin as a whole, high surface runoff did not necessarily mean high sediment yields, as is the case with wheatgrass. Similarly, mixed forests with the lowest yields (6.5 t km^{-2} yr^{-1}) have only the second-lowest surface runoff rate (66 mm yr^{-1}) (Figure 8b). To put Figure 8a and b in perspective, we plotted the average loads produced by each land use in Figure 8c. From the plot, pastureland, which otherwise has a very high specific yield, actually produced one of the least loads commensurate with its overall percentage of basin area. Likewise for agriculture, mixed forest, and non-forested wetlands.

4.1.3. Sediment Source Areas

Erodibility values calculated with the Kuery 1.4 model are compared with the original SWAT default values in Figure 9a,b The resulting soil erodibility values were a magnitude of 10 less than those estimated by default SWAT values. This is in line with [62] Salvador Sanchis et al. (2008), who showed that soils in warm climates are less erodible than those developed in temperate climates. The map of slope steepness (Figure 9c) also affirms that higher slopes will yield more sediment. Our resulting mapped values compared well with literature, notably the work of [17] Mushi et al. (2019), who used the Global Assessment of Soil Degradation (GLASOD) to map sediment sources within the CRB. Our simulated sediment sources (Figure 9d) is similar to theirs. The results showed that portions of the middle CRB are less affected by erosion, while high relief areas are the primary sources of sediment. It is important to note also the Cuvette central area, although located on low slopes with low erodible soils, yield relatively high sediment within its tributaries because of their high content in organic matter, which are mostly made up of fine particles that are transported easily even at low velocities. The coarser particles, most likely from stream bank erosion, will travel only shorter distances and be deposited, helping to create anastomosing structures observed along the middle reach [83,87,88] (O'Loughlin 2013, 2020; Carr et al., 2019). Thus, a combination of higher rainfall and corresponding surface runoff induced more sediment yield from the Cuvette Centrale. These results would need a thorough field-based verification to compensate for the uncertainties associated with the model, and as [17] Mushi et al. (2019) admitted, the GLASOD map is unable to map all the erosion processes.

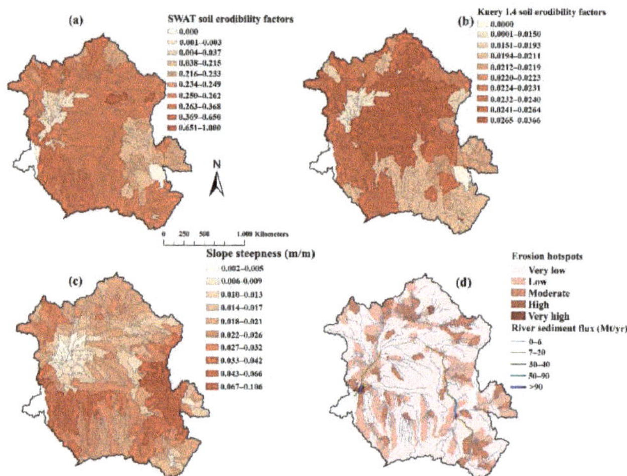

Figure 9. Maps over the Congo River basin comparing the K$_{usle}$ factor calculated with (**a**) SWAT and (**b**) Kuery 1.4, (**c**) slope steepness across the basin and (**d**) areas more prone to soil erosion within the simulation period.

4.2. Key Factors Affecting Sediment Yield

A principal component analysis (PCA) was performed on parameters that affect the sediment yield in the various subbasins of the catchment. Five key parameters were analysed: the slope in the HRUs (HRU_slp), precipitation (PCP), soil erodibility (K_{USLE}), channel vegetation cover (CH_COV2), and surface runoff (SUR_Q). PC1 explained 33.7% of the data variability, while PC2 explained 24.4%. Together, PCA1 and PCA2 accounted for 56.1% of the variability in the data (Table 7). The eigenvalues with the most magnitude in PC1 were 0.66 and 0.58, corresponding to surface runoff and precipitation, respectively. Figure 10a demonstrates this; as values increase positively along the PC1, so does precipitation and surface runoff. In addition, erodibility and slope are moderate and there is less channel cover. Similarly, the eigenvalues with the most magnitude in PC2 are erodibility and the slope in the HRUs with values of -0.62 and -0.59, respectively. Fig. 10b shows that as the values along with PC2 increase negatively, there is a corresponding increase in slope as well as the erodibility of the soils. At the same time, channel cover is low and there is no real difference between precipitation and surface runoff.

Table 7. PCA components and their respective eigenvalues and correlation coefficients.

	PC1	PC2	PC3	PC4	PC5	PC1	PC2
Proportion of variance (%)	33.7	22.4	19.9	16	8	Pearson coefficient	
Cumulative proportion of variance (%)	33.7	56.1	76	92	100		
			Eigenvalues				
HRU_SLP	0.35	−0.59	−0.30	−0.58	−0.31	0.46	−0.62
PCP	0.58	0.41	0.27	0.14	−0.64	0.75	0.43
_K_{USLE}	0.29	−0.62	0.11	0.71	0.07	0.38	−0.66
CH_COV2	−0.15	−0.26	0.90	−0.30	0.01	−0.20	−0.27
SUR_Q	0.66	0.17	0.09	−0.19	0.70	0.86	0.18

To buttress these results, we used the Pearson correlation coefficient to compare the relationship between the variables and our two principal components that explain more than 50% of the data variability. Table 7 shows the highest positive correlation along PC1 of 0.86 associated with surface runoff. This implies that as the value of the observation increases along PC1 so does the surface runoff. Likewise, the next highest value is precipitation (0.75), further reinforcing the dominance of these variables in the CRB sediment processes. The reverse is also true for negative coefficient values; as the value of the principal components increase, the associated variables decrease. Low negative values would also mean negligible relationships between variables as observed with channel vegetation cover in PC1 and PC2.

Our results compare favourably with the literature cited in this study [9,17,39,69,82] (Laraque et al., 2009; Runge et al., 2007; Vanmaercke et al., 2013; Mushi et al., 2019; Kabuya et al., 2020). The principal components can isolate the main factors responsible for the sediment transfer processes in the basin. The regions with relatively higher rainfall and surface runoff, which are the dominant factors along PC1, are situated in the central and highland areas of the east as well as portions of the southwest of the basin. While along PC2, the slope and the erodibility of the soils are closely related. These variables dominate the erosional processes in the highland regions, which surround the basin as well as parts of the main drainage units of the basin. The PCA is also able to group main drainage units with similar characteristics. By condensing our variables into two linear combinations of PC1 and PC2, it will be possible to separate our sub-basins based on the similarity of characteristics. Further work is therefore needed to separate the sub-basins based on the factors influencing erosion. This would need more detailed and extensive data coupled with modelling for a deeper understanding of the sediment dynamics within the CRB.

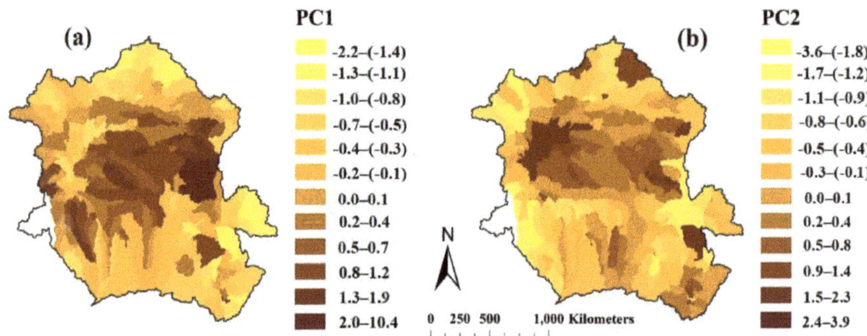

Figure 10. Principal component PC1 (**a**) and PC2 (**b**) values across the CRB simulated with the SWAT model.

4.3. Sediment Balance in the Cuvette Centrale

The Cuvette Centrale acts as a sediment trap slowing the flow of water, nutrients and pollutants [9] (Laraque et al., 2009). To quantify the effects of the Cuvette Centrale on upstream sediment loads, we carried out an upstream-downstream mass balance as described in Section 2.5 and as demonstrated in the hydrological balance of [50] Datok et al. (2021). Eight tributaries from both the northern and equatorial regions were considered and the results of the balance are presented in Table 8 below.

Table 8. Mean yearly simulated averages of suspended sediment fluxes and sediment yield for the 2000 to 2012 period.

Rivers	Contributing Area (km^2)	Station/Town	#	*Annual Load (ton)	% of Total Load Supplied to the Cuvette Centrale	Specific Yield (t km^{-2} yr^{-1})
Congo	1,306,000	Mbandaka	7	*26,981,769*	85.85	20.66
Ruki	163,700	Bokuma	8	*1,602,153*	5.10	9.79
Ubangi	491,800	Mongoumba	6	*1,502,746*	4.78	3.06
Sangha	138,600	Ouesso	4	*942,815*	3.0	6.80
Alima	18,240	Tchikapika	1	*187,469*	0.60	10.28
Kouyou	9631	Linnegue	2	*96,124*	0.30	9.98
Likouala aux Herbes	24,570	Botouali	5	*76,227*	0.24	3.10
Likouala Mossaka	12,240	Makoua	3	*40,495*	0.13	3.31
Congo	2,343,000	Bolobo	9	7,624,230		3.25
Kasaï	778,900	Kwamouth	10	10,503,792		13.49
Congo	3,164,000	Brazzaville	11	37,094,615		11.72

* Values in italics were summed for the total inputs.

We simulated an average of 31.4 million tons of particulate material filling the Cuvette Centrale annually ($I = \sum_{i=1}^{8} S_i$). A large percentage of the fluxes (85%) come from the upper Congo. The Ubangi, the largest tributary of the Congo on the North, contributes about five percent of fluxes while the Ruki, another large tributary with its source in the southern hemisphere also contributes about five percent (Table 8). The Sangha and the other principal tributaries on the right bank contribute three and two per cent to the Cuvette Centrale, respectively.

Compared to the values of 33.7×10^6 t yr^{-1} of material measured by the SO-HYBAM at the basin outlet (station 11) from 2006 to 2017, our simulated value of 37×10^6 t yr^{-1} for 2000 to 2012 compares well. Next, we calculated the fluxes at the confluence of the Kasai River (at Kwamouth; station 10) with the Congo at 10.5 million tons, which was then subtracted from the flux of 37 million tons simulated at the basin outlet resulting in 26.6 million tons of material annually at the basin outlet. With respect to equation 6 in Section 2.7, we estimated the sediment balance as the difference between the fluxes coming in from stations 1 to 8 (~31Mt) and the fluxes at station 11 (~26 Mt) for a balance of ~5 Mt. To verify the balance, we calculated the fluxes downstream of the Cuvette at Bolobo, which is 7.6 million tons of material. By subtracting the value at Bolobo from the inputs into the Cuvette Centrale gave a value of over 23 million tons, we confirmed that there is a massive reduction in the fluxes exiting the Cuvette Centrale. These high loads retained by the cuvette can be justified by the sediment yields simulated in Figure 7 as well as the anastomosing structures with low velocities (0.75–0.95 m s^{-1}) observed by [83] Carr et al. (2019) which are conducive for sediment storage. [89] Molliex et al. (2019) also suggested that the Cuvette Centrale, including the lakes Tumba and Mai-Ndombe, trap a huge part of the CRB sediment load. [90] Garzanti et al. (2019) from petrographic evidence, suggests that pure quartzite is generated only in the Cuvette Centrale and that the Cuvette Centrale is a quartz factory with a huge reservoir. They also estimated that the Congo is the largest River on Earth carrying pure quartzite to the Atlantic. Coynel et al. (2005) estimated that more than 11 megatons of suspended sediments are exported from only the Savannah areas of the CRB with around 23% of that amount unaccounted for at the basin outlet. While there is no substitute for field-based verification, these studies suggest that the Cuvette Centrale can store such amounts.

Therefore, for the period 2000 to 2012, we estimated that the Cuvette Centrale retained over 23 megatons of material or 75% of the sediment fluxes from 70% of the basin area above Brazzaville/Kinshasa (combined contributing areas of stations 1 to 8). The low velocities observed in this area of the wetlands [83,91] (Laraque et al., 1998a; Carr et al., 2019) would ensure deposition takes place. This would imply a minimum exchange of material between the wetland and the main river. [50] Datok et al. (2021) analysed the water balance of the Cuvette Centrale and verified its role as a sponge or buffer that accumulates water in the rainy season and releases it in the dry season. A running hypothesis suggests that this may be the main means by which supplies–if any–of suspended sediment is made to the main river [33] (Alsdorf et al., 2016). Dilution and concentration effects have also been observed at the peak and low of discharge, respectively [27] (Moukandi N'kaya et al., 2020). This is expected as it has been observed that increasing rainfall increases the specific discharge causing dilution of suspended matter [9] (Laraque et al., 2009). [91] Laraque et al. (1998a) described the hydrological functioning of the floodable parts of the Cuvette Centrale as mainly through vertical exchanges. Thus, coupled with the flat topography and limited lateral exchanges, there is poor mineralization. These characteristics can also be extended to the Lake regions on the left bank of the Congo River i.e., Lakes Tumba and Mai Ndombe, which are located within seasonally inundated forests. They are reservoirs of organic material due to the forest cover and their biogeochemistry is similar to the flooded portions of the Cuvette Centrale.

The seasonal dynamics of the loads contributed to the Cuvette Centrale (Figure 11) show that the right bank tributaries (stations 1–6; Table 8) contribute more load during the September-October-November (SON) floods associated with the equatorial regions. Disregarding the contributions from the main-stem Congo River (station 7); it is obvious that the right bank tributaries are a very important source of sediment to the Cuvette Centrale. This is especially so since unlike the Amazon which has numerous connecting pathways or "furos" between the main river and the floodplains, there is limited connectivity in the case of the CRB [33] (Alsdorf et al., 2016). The left bank tributary (station 8) also makes a modest contribution in the December-January-February (DJF) period that coincides with the small dry season of the southern basins and the pronounced dry season of the northern basins.

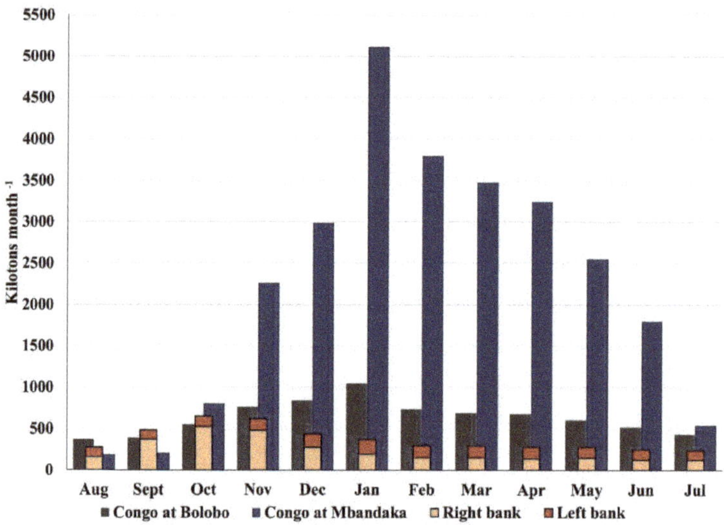

Figure 11. Mean interannual seasonal dynamics of the loads contributed to the Cuvette Centrale from tributaries on the right (Alima, Kouyou, Likouala Mossaka, Sangha, Likouala aux Herbes, Ubangi) and left (Ruki) of the main river, as well as the loads coming in and carried away from the area (Congo at Mbandaka and Bolobo, respectively).

4.4. Comparison with the Amazon and the Orinoco

To put this study in context and appreciate the functioning of the CRB, we compare the material fluxes of the CRB with two other large world drainage basins: the Amazon and the Orinoco. This comparison is important because, after the Amazon, the Congo is the second largest area of tropical rainforest at 1.8 million km^2 [92] (Haensler et al., 2013); and its average discharge of 36,150 m^3 s^{-1} (this study) is second to the Amazon with 169,000 m^3 s^{-1} [93] (Filizola & Guyot, 2009) (Table 9). In comparison to the Orinoco, their mean annual discharges are similar, with the Orinoco discharging 31,400 m^3 s^{-1}. Together, the three river basins combine an average discharge of approximately 284 × 10^3 m^3 s^{-1} of water into the Atlantic [43] (Laraque et al., 2013). The sediment fluxes reported in Table 9 for the Amazon are taken from the Obidos station that controls 80% of the basin. In terms of the fluxes, both the Amazon and the Orinoco have much higher amounts discharged to the oceans. This is principally due to the difference in relief compared to the Congo as the South American rivers are underlain in part by the Andes [94] (Meade, 1984). In the Congo, chemical weathering dominates over physical weathering with dissolved matter superseding suspended sediments [27,43] (Laraque et al., 2013a; Moukandi N'kaya et al., 2020). The reverse is the case for the Orinoco; due to the influence of mountain chains on discharge and fluxes, the differences in the hydrological regimes, forest cover, and the high seasonal variability of discharge compared to the Congo [43,95] (Richey et al., 1986; Laraque et al., 2013a).

Table 9. Comparison of the discharge and Suspended sediment fluxes in three of the world's largest river basins.

	Congo at BRZ/KIN (This Study)	Orinoco at Ciudad Bolivar (Laraque et al., 2013a [43])	Amazon at Obidos (Filizola & Guyot, 2009 [93])
Q (m^3 s^{-1})	36,150	31,400	169,000
SSC (mg L^{-1})	31.48	73.9	-
SS load (10^6 t yr^{-1})	37.0	74.0	555
SS yield (t km^{-2} yr^{-1})	11.72	88.5	120

4.5. Uncertainties Related to Sediment Estimation

As with any large scale hydrological modelling study in ungauged basins, uncertainties are unavoidable. In this study, a lack of critical data for calibrating the sediment fluxes caused the model to be calibrated in a lumped manner. There is an unavoidable loss of information caused by the lack of spatial representation. However, the temporal variability was well captured and the magnitude of fluxes matched the literature. It should also be noted that concerns have been raised about the misrepresentation of wetland processes due to gaps in the model structure [50] (Datok et al., 2021). Therefore, this study was undertaken assuming that the processes occurring in the wetlands are straightforward. There is also uncertainty related to input data. For example, the soil database used to construct the SWAT lookup tables was primarily prepared for agricultural use with soil depths not exceeding 100 cm. Therefore, peak flows could be over-simulated in deeper soils resulting in an overestimation of sediment. The DEM resolution for computing the hillslope length also has implications for runoff and erosion computations. The Bagnold equation used to calculate the maximum amount of sediment transported in a stream segment has its limitations [96] (Wei et al., 2020). The algorithm does not differentiate between particle sizes and offers no distinction between channel and flood plain deposition [49] (Neitsch et al., 2011). Precipitation input uncertainty and the uncertainties of parameters of the central basin all have to be taken into account too.

In estimating the sediment balance of the Cuvette Centrale, an analysis of stream hydraulics will also help in determining if the widths and depths of the tributaries studied can process the amounts of sediment over the period considered. Settling rates of sediment in the Cuvette Centrale will also need to be studied. This is important in order not to overlook the possibility that the Cuvette Centrale is subsiding at a rate equivalent to the trapping of the sediments. This is an area of future work that will need exploring.

The existing dams that could affect our results are the Mobayi Dam downstream of the Uele River with an installed capacity of 11.5 MW and the Imboulou Dam on the Lefini River with an installed capacity of 120 MW. In the case of the former, it is upstream of the Bangui station and for the latter, it is downstream of the Cuvette Centrale. The Lualaba basin has swamps and lakes that serve as natural reservoirs with a few man-made ones. However, reservoirs were also integrated into the model to account for these structures. Considering the scale of the study, these effects although negligible, will be a source of uncertainty.

5. Conclusions

Understanding the sediment dynamics in wetlands is important to assess changes taking place over time correctly. These changes could be natural or human-made and could have important ecological as well as socio-economic consequences. We used the SWAT hydrological model in this study to investigate and quantify sediment fluxes and budgets from seasonal to annual time scales using freely available datasets. The SWAT model satisfactorily captured the dynamics of sediment yields and fluxes when compared to past studies. In summary, we have been able to use the SWAT model to do the following:

I. Run the SWAT model on a daily timescale, calibrating the hydrology and sediment dynamics for the 2000 to 2012 period. Five principal catchments of the CRB were calibrated for hydrology based on previous work [50] (Datok et al., 2021). Due to lack of data, we were only able to calibrate one station in the upper part of the basin as well as the outlet of the basin for the sediments. We achieved acceptable results based on the performance of the evaluation criteria used in the study. For hydrology, the coefficients of efficiency were above 70% in 60% of the sub-basins calibrated, while other evaluation criteria also gave a good picture of the model performance.

II.   An assessment of the model outputs revealed that agriculture and pasture were the main land uses with the potential to contribute the most sediment yield. The absence of any major infrastructure like dams in most of the tributaries, to trap sediment even temporarily, makes this an important source of sediments. Generally, the steepest areas of the basin produced more sediment and runoff. Our map of degradation hotspots also revealed that the middle basin is least affected by erosion with degradation more pronounced in the outer rim of the Cuvette Centrale. At the same time, the Cuvette Centrale area showed the lowest values of soil erodibility mainly because of the soil properties. Thus, the erosion within this area is mainly local with corresponding sediment accumulation. The highest specific sediment yields by sub-basin were from the Kasai, aided by intense anthropogenic disturbances that lead to deforestation and degradation of uncovered land. The Ubangi subbasin, on the other hand, an old erosional surface, has the lowest specific sediment yields. In terms of load carried in stream channels, the heaviest loads were carried from the lakes at the southeastern parts of the basin and the confluence of the Kasai with the main Congo River.

III.   Principal component analysis revealed the main factors that aid in sediment transport. They include surface runoff, precipitation, the slope in the HRU and the erodibility of the soils. These factors are closely related to the climate and physiography of the catchment with high relief areas more prone to erosion and the high precipitation causing increased surface runoff in the central basin.

We estimated the balance of fluxes entering and exiting the Cuvette Centrale from eight principal tributaries that contribute to the main river. Our estimates suggest that the Cuvette Centrale is indeed a giant reservoir that retains at least 23 megatons of material annually. Sediment from the upper Congo is the main contributor with 85% of the fluxes, while the principal tributary of the right bank- the Ubangi- contributes 5%. The Ruki also contributes about 5%, while the other right bank affluents contribute the remainder. With the lack of a robust database over the entire basin, the results over those areas were measured data are available, provide a higher degree of confidence. More advanced validations are needed over the ungauged portions especially at Mbandaka station as the upper Congo is the main supplier of sediments to the Cuvette Centrale.

This work also showed that the estimation of soil erodibility should not only depend on lithological information or a geological map. This is because their reliability is variable and could lead to over or under-estimates of the K values. Future improvements to the model will focus on the spatial representativeness of the sedimentation processes as well as the use of high-resolution input data. Incorporation of a wetland module will also aid in a proper understanding of the processes in the central ungauged basin (e.g., [97–99] Hughes et al., 2014; Sun et al., 2016, 2018).

This study is a follow-up to [50] Datok et al., (2021), where a water balance of the Cuvette Centrale was made. They found that as much water passed through the Cuvette Centrale from the upper Congo as efficient precipitation falling over the area. Similarly, we estimated that a majority of the fluxes come from the upper Congo. The implications are that sediment, nutrients and pollutants are supplied to the Cuvette Centrale provided there are connecting pathways or overflows from the banks [33] (Alsdorf et al., 2016). The relatively higher water surface slopes in the middle reach passing through the Cuvette Centrale make this a real possibility. Sediment loads and yields are majorly influenced by land cover and land use, which alters upland erosion processes [96,100] (Wei et al., 2020; Sok et al., 2020). Rapid urbanization, hydroelectric dam construction and/or future water diversion schemes would also have a significant effect on the fluxes of material in the Cuvette Centrale. The consequences on downstream communities of wetland loss and nutrient imbalance from sedimentation are severe and should be given more attention.

For the first time, we have estimated the fluxes trapped by the Cuvette Centrale using an upstream-downstream mass balance approach. Our model can be used as a baseline for future sediment studies of the CRB. Different modelling approaches can also be initiated for the Cuvette Centrale from the point scale to the macro-level and vice versa to understand the processes operating within. Concerted joint efforts by stakeholders are also required if progress is to be made in fully appreciating the functioning of this still, relatively pristine watershed.

Author Contributions: Conceptualization, P.D., S.S. and J.-M.S.-P.; methodology, P.D., S.S. and J.-M.S.-P.; software, P.D., S.S., C.F. and J.-M.S.-P.; validation, P.D., S.S., C.F., A.L., S.O. and J.-M.S.-P.; formal analysis, P.D., S.S., C.F., A.L., S.O. and J.-M.S.-P.; investigation, P.D., S.S., C.F., A.L., S.O. and J.-M.S.-P.; resources, P.D., S.S., C.F., A.L., S.O., G.M.N. and J.-M.S.-P.; data curation, P.D., A.L. and G.M.N.; writing–original draft preparation, P.D.; writing–review and editing, P.D., S.S., C.F., A.L., S.O. and J.-M.S.-P.; visualization, P.D., S.S., C.F., A.L., S.O., G.M.N. and J.-M.S.-P.; supervision, P.D., S.S. and J.-M.S.-P. All authors have read and agreed to the published version of the manuscript.

Funding: This research received no external funding.

Institutional Review Board Statement: Not applicable.

Informed Consent Statement: Not applicable.

Acknowledgments: The first Author wishes to thank the TETFUND for funding his stay at the Laboratoire Ecologie Fonctionnelle et Environnement, Université de Toulouse.

Conflicts of Interest: The authors declare no conflict of interest.

References

1. Hay, W.W. Detrital sediment fluxes from continents to oceans. *Chem. Geol.* **1998**, *145*, 287–323. [CrossRef]
2. Walling, D. Human impact on land–ocean sediment transfer by the world's rivers. *Geomorphology* **2006**, *79*, 192–216. [CrossRef]
3. Haddadchi, A.; Ryder, D.; Evrard, O.; Olley, J. Sediment fingerprinting in fluvial systems: Review of tracers, sediment sources and mixing models. *Int. J. Sediment Res.* **2013**, *28*, 560–578. [CrossRef]
4. NKounkou, R.-R.; Probst, J.-L. Hydrology and geochemistry of the Congo river system. *SCOPE/UNEP-Sonderband* **1987**, *64*, 483–508.
5. Walling, D.E. *The Impact of Global Change on Erosion and Sediment Transport by Rivers: Side Publications Series The United Nations World Water Development Report 3*; United Nations World Water Assessment Program: Paris, France, 2009; Volume 81, p. 26.
6. Mitsch, W.J.; Gosselink, J.G. *Wetlands*, 5th ed.; John Wiley & Sons, Inc.: Hoboken, NJ, USA, 2015.
7. Gosselink, J.G.; Turner, R.E. The role of hydrology in freshwater wetland ecosystems. In *Freshwater Wetlands: Ecologicalprocesses and Management Potential*; Good, R.E., Whigham, D.F., Simpson, R.L., Eds.; Academic Press: New York, NY, USA, 1978; pp. 63–78.
8. Novitzki, R.P. Hydrologic characteristics of Wisconsin's wet-lands and their influence on floods, stream flow, and sediment. In *Wet-Land Functions and Values: The State of Our Understanding*; Greeson, P.E., Clark, J.R., Clark, J.E., Eds.; American Water Resources Association: Minneapolis, MN, USA, 1979; pp. 377–388.
9. Laraque, A.; Bricquet, J.P.; Pandi, A.; Olivry, J.C. A review of material transport by the Congo River and its tributaries. *Hydrol. Process.* **2009**, *23*, 3216–3224. [CrossRef]
10. Mitsch, W.J.; Bernal, B.; Hernandez, M.E. Ecosystem services of wetlands. *Int. J. Biodivers. Sci. Ecosyst. Serv. Manag.* **2015**, *11*, 1–4. [CrossRef]
11. Dargie, G.C.; Lewis, S.L.; Lawson, I.T.; Mitchard, E.T.A.; Page, S.E.; Bocko, Y.E.; Ifo, S.A. Age, extent and carbon storage of the central Congo Basin peatland complex. *Nat. Cell Biol.* **2017**, *542*, 86–90. [CrossRef] [PubMed]
12. Harenda, K.M.; Lamentowicz, M.; Samson, M.; Chojnicki, B.H. The Role of Peatlands and Their Carbon Storage Function in the Context of Climate Change. In *Interdisciplinary Approaches for Sustainable Development Goals*; Springer: Cham, Switzerland, 2018; pp. 169–187. [CrossRef]
13. Moukolo, N.; Laraque, A.; Olivry, J.C.; Bricquet, J.P. Transport en solution et en suspension par le fleuve Congo (Zaïre) et ses principaux affluents de la rive droite. *Hydrol. Sci. J.* **1993**, *38*, 133–145. [CrossRef]
14. Gaillardet, J.; Dupré, B.; Allègre, C.J. A global geochemical mass budget applied to the Congo basin rivers: Erosion rates and continental crust composition. *Geochim. Cosmochim. Acta* **1995**, *59*, 3469–3485. [CrossRef]
15. Dupré, B.; Gaillardet, J.; Rousseau, D.; Allègre, C.J. Major and trace elements of river-borne material: The Congo Basin. *Geochim. Cosmochim. Acta* **1996**, *60*, 1301–1321. [CrossRef]
16. Guyot, J.L.; Olivry, J.C.; Laraque, A.; Orange, D. Discharge and Suspended Sediment Fluxes in Large Tropical Rivers. In Proceedings of the IGBP Synthesis Workshop, Stockholm, Sweden, 1 February 2000; pp. 1992–1994.

17. Mushi, C.; Ndomba, P.; Trigg, M.; Tshimanga, R.; Mtalo, F. Assessment of basin-scale soil erosion within the Congo River Basin: A review. *Catena* **2019**, *178*, 64–76. [CrossRef]

18. Probst, J.L. Carbon River Flues and Weathering CO_2 Consumption in the Congo and Amazon River Basins. *Appl. Geochem.* **1983**, *9*, 1–13. [CrossRef]

19. Probst, J.-L.; Nkounkou, R.-R.; Krempp, G.; Bricquet, J.-P.; Thiébaux, J.-P.; Olivry, J.-C. Dissolved major elements exported by the Congo and the Ubangi rivers during the period 1987–1989. *J. Hydrol.* **1992**, *135*, 237–257. [CrossRef]

20. Seyler, P.; Coynel, A.; Moreira-Turcq, P.; Etcheber, H.; Colas, C.; Orange, D.; Bricquet, J.-P.; Laraque, A.; Guyot, J.L.; Olivry, J.-C.; et al. Organic Carbon Transported by the Equatorial Rivers: Example of Congo-Zaire and Amazon Basins. *Soil Eros. Carbon Dyn.* **2005**, 255–274. [CrossRef]

21. Gumbricht, T.; Roman-Cuesta, R.M.; Verchot, L.; Herold, M.; Wittmann, F.; Householder, E.; Herold, N.; Murdiyarso, D. An expert system model for mapping tropical wetlands and peatlands reveals South America as the largest contributor. *Glob. Chang. Biol.* **2017**, *23*, 3581–3599. [CrossRef] [PubMed]

22. Nguimalet, C.-R.; Orange, D. Caractérisation de la baisse hydrologique actuelle de la rivière Oubangui à Bangui, République Centrafricaine. *Houille Blanche* **2019**, 78–84. [CrossRef]

23. Aagaard, T.; Hughes, M. *10.4 Sediment Transport*; Elsevier BV: Amsterdam, The Netherlands, 2013; Volume 10, pp. 74–105.

24. Oeurng, C.; Sauvage, S.; Coynel, A.; Maneux, E.; Etcheber, H.; Sánchez-Pérez, J.-M. Fluvial transport of suspended sediment and organic carbon during flood events in a large agricultural catchment in southwest France. *Hydrol. Process.* **2011**, *25*, 2365–2378. [CrossRef]

25. Guan, Z.; Tang, X.-Y.; Yang, J.E.; Ok, Y.S.; Xu, Z.; Nishimura, T.; Reid, B.J. A review of source tracking techniques for fine sediment within a catchment. *Environ. Geochem. Health* **2017**, *39*, 1221–1243. [CrossRef] [PubMed]

26. Laraque, A.; N'Kaya, G.D.M.; Orange, D.; Tshimanga, R.; Tshitenge, J.-M.; Mahé, G.; Nguimalet, C.R.; Trigg, M.A.; Yepez, S.; Gulemvuga, G. Recent Budget of Hydroclimatology and Hydrosedimentology of the Congo River in Central Africa. *Water* **2020**, *12*, 2613. [CrossRef]

27. Moukandi, G.D.N.; Orange, D.; Murielle, S.; Padou, B.; Datok, P.; Laraque, A. Temporal Variability of Sediments, Dissolved Solids and Dissolved Organic Matter Fluxes in the Congo River at Brazzaville/Kinshasa. *Geosciences* **2020**, *10*, 341.

28. Moukandi, G.D.N.; Laraque, A.; Paturel, J.; Gulemvuga, G.; Mahé, G. A New Look at Hydrology in the Congo Basin, Based on the Study of Multi-Decadal Chronicles. In *Congo Basin Hydrology, Climate, and Biogeochemistry: A Foundation for the Future*; Geophysical Monograph Series; Alsdorf, D., Moukandi, G., Tshimanga, R., Eds.; John Wiley & Sons Inc.: Hoboken, NJ, USA, 2021. [CrossRef]

29. BRLi. Développement et mise en place de l'outil de modélisation et d'allocation des ressources en eau du Bassin du Congo. In *Rapport Technique de Construction et de Calage du Modèle*; CICOS: Kinshasa, Republic of the Congo, 2016.

30. Kadima, E.; Delvaux, D.; Sebagenzi, S.N.; Tack, L.; Kabeya, S.M. Structure and geological history of the Congo Basin: An integrated interpretation of gravity, magnetic and reflection seismic data. *Basin Res.* **2011**, *23*, 499–527. [CrossRef]

31. Daly, M.C.; Lawrence, S.R.; Diemu-Tshiband, K.; Matouana, B. Tectonic evolution of the Cuvette Centrale, Zaire. *J. Geol. Soc.* **1992**, *149*, 539–546. [CrossRef]

32. Laraque, A.; Mietton, M.; Olivry, J.C.; Pandi, A. Influence Des Couvertures Lithologiques et Végétales Sur Les Régimes et La Qualité Des Eaux. *J. Water Sci.* **1998**, *11*, 209–224.

33. Alsdorf, D.; Beighley, E.; Laraque, A.; Lee, H.; Tshimanga, R.; O'Loughlin, F.; Mahé, G.; Dinga, B.; Moukandi, G.; Spencer, R.G.M. Opportunities for hydrologic research in the Congo Basin. *Rev. Geophys.* **2016**, *54*, 378–409. [CrossRef]

34. Bwangoy, J.-R.B.; Hansen, M.C.; Roy, D.P.; De Grandi, G.; Justice, C.O. Wetland mapping in the Congo Basin using optical and radar remotely sensed data and derived topographical indices. *Remote Sens. Environ.* **2010**, *114*, 73–86. [CrossRef]

35. Lehner, B.; Döll, P. Development and validation of a global database of lakes, reservoirs and wetlands. *J. Hydrol.* **2004**, *296*, 1–22. [CrossRef]

36. Crosby, A.G.; Fishwick, S.; White, N. Structure and evolution of the intracratonic Congo Basin. *Geochem. Geophys. Geosystems* **2010**, *11*, 1–20. [CrossRef]

37. UNESCO. Resources of the World and Their Use. 2004. Available online: https://unesdoc.unesco.org/ark:/48223/pf0000134433 (accessed on 4 February 2021).

38. Le Heron, D.P. *Schlüter, T. Geological Atlas of Africa, with Notes on Stratigraphy, Economic Geology, Geohazards and Geosites of Each Country*; Springer: Berlin/Heidelberg, Germany, 2006. [CrossRef]

39. Kabuya, P.M.; Hughes, D.A.; Tshimanga, R.M.; Trigg, M.A.; Bates, P. Establishing uncertainty ranges of hydrologic indices across climate and physiographic regions of the Congo River Basin. *J. Hydrol. Reg. Stud.* **2020**, *30*, 100710. [CrossRef]

40. Olivry, J.-C.; Bricquet, J.-P.; Thiébaux, J.-P.; Sigha-Nkamdjou, L. Transport de matière sur les grands fleuves des régions intertropicales: Les premiers résultats des mesures de flux particulaires sur le bassin du fleuve Congo. In *Sediment Budgets*; AISH: Porto-Alegre, Brazil, 1988; pp. 509–521.

41. Olivry, J.C.; Bricquet, J.P.; Laraque, A.; Guyot, J.L. Flux Liquides, Dissous et Particulaires de Deux Grands Bassins Intertropicaux: Le Congo à Brazzaville et Le Rio Madeira à Villabella. In *Grands Bassins Fluviaux Périatlantiques: Congo, Niger, Amazone*; Olivry, J.C., Boulegue, J., Eds.; IRD: Marseille, France, 1995; pp. 345–355.

42. Coynel, A.; Seyler, P.; Etcheber, H.; Meybeck, M.; Orange, D. Spatial and seasonal dynamics of total suspended sediment and organic carbon species in the Congo River. *Glob. Biogeochem. Cycles* **2005**, *19*, 1–17. [CrossRef]

43. Laraque, A.; Castellanos, B.; Steiger, J.; Lòpez, J.L.; Pandi, A.; Rodriguez, M.; Rosales, J.; Adèle, G.; Perez, J.; Lagane, C. A comparison of the suspended and dissolved matter dynamics of two large inter-tropical rivers draining into the Atlantic Ocean: The Congo and the Orinoco. *Hydrol. Process.* **2013**, *27*, 2153–2170. [CrossRef]

44. Wechsler, S.P. Uncertainties associated with digital elevation models for hydrologic applications: A review. *Hydrol. Earth Syst. Sci.* **2007**, *11*, 1481–1500. [CrossRef]

45. Nachtergaele, F.A.; van Velthuizen, H.B.; Batjes, N.C.; Dijkshoorn, K.C.; van Engelen, V.C.; Fischer, G.B.; Jones, A.D.; Montanarella, L.D.; Petri, M.A.; Prieler, S.B.; et al. The Harmonized World Soil Database Food and Agriculture Organization of the United Nations. In Proceedings of the World Congress of Soil Science, Soil Solutions for a Changing World, Brisbane, Australia, 1–6 August 2010; Published on DVD. pp. 34–37.

46. Huffman, G.J.; Bolvin, D.T.; Nelkin, E.J.; Wolff, D.B.; Adler, R.F.; Gu, G.; Hong, Y.; Bowman, K.P.; Stocker, E.F. The TRMM Multisatellite Precipitation Analysis (TMPA): Quasi-Global, Multiyear, Combined-Sensor Precipitation Estimates at Fine Scales. *J. Hydrometeorol.* **2007**, *8*, 38–55. [CrossRef]

47. Hartmann, J.; Moosdorf, N. The new global lithological map database GLiM: A representation of rock properties at the Earth surface. *Geochem. Geophys. Geosystems* **2012**, *13*, 1–37. [CrossRef]

48. Shekhar, S.; Xiong, H. Soil and Water Assessment Tool "SWAT". *Texas Water Resour. Inst.* **2008**, 1068. [CrossRef]

49. Zhou, P.; Huang, J.; Pontius, R.G.; Hong, H. New insight into the correlations between land use and water quality in a coastal watershed of China: Does point source pollution weaken it? *Sci. Total Environ.* **2016**, *543*, 591–600. [CrossRef] [PubMed]

50. Datok, P.; Fabre, C.; Sauvage, S.; N'kaya, G.M.; Paris, A.; Dos Santos, V.; Laraque, A.; Sánchez Pérez, J.-M. Investigating the role of the Cuvette Centrale in the hydrology of the Congo River Basin. In *Congo Basin Hydrology, Climate, and Biogeochemistry: A Foundation for the Future*; Geophysical Monograph Series; Alsdorf, D., Moukandi, G., Tshimanga, R., Eds.; John Wiley & Sons Inc.: Hoboken, NJ, USA, 2021. [CrossRef]

51. Gupta, H.V.; Kling, H.; Yilmaz, K.K.; Martinez, G.F. Decomposition of the mean squared error and NSE performance criteria: Implications for improving hydrological modelling. *J. Hydrol.* **2009**, *377*, 80–91. [CrossRef]

52. Williams, J.R. Sediment yield prediction with universal equation using runoff energy factor. In *Present and Prospective Technology for Predicting Sediment Yields and Sources, Proceedings of the Sediment-Yield Workshop, USDA Sedimentation Laboratory, Oxford, MI, USA, 17–18 January 1973*; U.S. Department of Agriculture, Agricultural Research Service: Washington, DC, USA, 1975; pp. 244–252.

53. Wischmeier, W.H.; Smith, D.D. *Predicting Rainfall-Erosion Losses: A Guide to Conservation Planning*; USDA Agriculture Handbook, Department of Agriculture, Science and Education Administration: Washington, DC, USA, 1978; p. 537.

54. Wischmeier, W.H.; Smith, D.D. *Predicting Rainfall-Erosion Losses from Croplands East of the Rocky Mountains*; USDA Agriculture Handbook; Department of Agriculture, Science and Education Administration: Washington, DC, USA, 1965; Volume 282, p. 47.

55. Vigiak, O.; Malagó, A.; Bouraoui, F.; Vanmaercke, M.; Poesen, J. Adapting SWAT hillslope erosion model to predict sediment concentrations and yields in large Basins. *Sci. Total Environ.* **2015**, *538*, 855–875. [CrossRef]

56. Jha, M.; Gassman, P.W.; Secchi, S.; Gu, R.; Arnold, J. Effect of watershed subdivision on swat flow, sediment, and nutrient predictions. *JAWRA* **2004**, *40*, 811–825. [CrossRef]

57. Borselli, L.; Cassi, P.; Sanchis, P. Soil Erodibility Assessment for Applications at Watershed Scale. In *Manual of Methods for Soil and Land Evaluation*; Science Publishers: Enfield, NH, USA, 2009. [CrossRef]

58. Borselli, L.; Torri, D.; Poesen, J.; Iaquinta, P. A robust algorithm for estimating soil erodibility in different climates. *Catena* **2012**, *97*, 85–94. [CrossRef]

59. Torri, D.; Poesen, J. Erodibilità. In *Metodi di Analisi Fisica del Suolo*; Pagliai, M., Ed.; Sezione VII; Ministero Delle Politiche Agricole e Forestali-Societa Italiana di Scienza del Suolo: Rome, Italy, 1997.

60. Torri, D.; Poesen, J.; Borselli, L. Corrigendum to "Predictability and uncertainty of the soil erodibility factor using a global dataset" [Catena 31 (1997) 1–22] and to "Erratum to Predictability and uncertainty of the soil erodibility factor using a global dataset" [Catena 32 (1998) 307–308]. *Catena* **2002**, *46*, 309–310. [CrossRef]

61. Shirazi, M.A.; Boersma, L.; Hart, J.W. A Unifying Quantitative Analysis of Soil Texture: Improvement of Precision and Extension of Scale. *Soil Sci. Soc. Am. J.* **1988**, *52*, 181–190. [CrossRef]

62. Sanchis, M.P.S.; Torri, D.; Borselli, L.; Poesen, J. Climate effects on soil erodibility. *Earth Surf. Process. Landf.* **2007**, *33*, 1082–1097. [CrossRef]

63. Engelen, V.; Verdoodt, A.; Dijkshoorn, J.A.; Van Ranst, E. *Soil and Terrain Database of Central Africa (DR of Congo, Burundi and Rwanda)*; Report 2006/07; ISRIC—World Soil Information: Wageningen, The Netherlands, 2006; Available online: http://www.isric.org (accessed on 13 May 2021).

64. Peel, M.C.; Finlayson, B.L.; McMahon, T.A. Updated world map of the Köppen-Geiger climate classification. *Hydrol. Earth Syst. Sci.* **2007**, *11*, 1633–1644. [CrossRef]

65. Bouillon, S.; Yambélé, A.; Spencer, R.G.M.; Gillikin, D.P.; Hernes, P.J.; Six, J.; Merckx, R.; Borges, A. Organic matter sources, fluxes and greenhouse gas exchange in the Oubangui River (Congo River basin). *Biogeosciences* **2012**, *9*, 2045–2062. [CrossRef]

66. Bouillon, S.; Yambélé, A.; Gillikin, D.P.; Teodoru, C.; Darchambeau, F.; Lambert, T.; Borges, A.V. Contrasting biogeochemical characteristics of the Oubangui River and tributaries (Congo River basin). *Sci. Rep.* **2014**, *4*, 5402. [CrossRef] [PubMed]

67. Laraque, A.; Mahé, G.; Orange, D.; Marieu, B. Spatiotemporal variations in hydrological regimes within Central Africa during the XXth century. *J. Hydrol.* **2001**, *245*, 104–117. [CrossRef]

68. Laraque, A.; Bellanger, M.; Adele, G.; Guebanda, S.; Gulemvuga, G.; Pandi, A.; Paturel, J.E.; Robert, A.; Tathy, J.P.; Yambele, A. Evolutions Récentes Des Débits Du Congo, de l'Oubangui et de La Sangha. *Geo. Eco. Trop.* **2013**, *37*, 93–100.
69. Vanmaercke, M.; Poesen, J.; Broeckx, J.; Nyssen, J. Sediment yield in Africa. *Earth Sci. Rev.* **2014**, *136*, 350–368. [CrossRef]
70. Moriasi, D.N.; Gitau, M.W.; Pai, N.; Daggupati, P. Hydrologic and Water Quality Models: Performance Measures and Evaluation Criteria. *Trans. ASABE* **2015**, *58*, 1763–1785. [CrossRef]
71. Adla, S.; Tripathi, S.; Disse, M. Can We Calibrate a Daily Time-Step Hydrological Model Using Monthly Time-Step Discharge Data? *Water* **2019**, *11*, 1750. [CrossRef]
72. Chishugi, J.B.; Alemaw, B.F. The Hydrology of the Congo River Basin: A GIS-Based Hydrological Water Balance Model. In *Proceedings of the World Environmental and Water Resources Congress*; Starett, S., Ed.; American Society of Civil Engineers (ASCE): Reston, VA, USA, 2009; pp. 1–16.
73. Tshimanga, R.M. Hydrological Uncertainty Analysis and Scenario-Based Streamflow Modelling for the Congo River Basin. Ph.D. Thesis, Rhodes University, Makhanda, South Africa, February 2012.
74. Tshimanga, R.M.; Hughes, D.A. Basin-scale performance of a semidistributed rainfall-runoff model for hydrological predictions and water resources assessment of large rivers: The Congo River. *Water Resour. Res.* **2014**, *50*, 1174–1188. [CrossRef]
75. Munzimi, Y.A.; Hansen, M.C.; Asante, K.O. Estimating daily streamflow in the Congo Basin using satellite-derived data and a semi-distributed hydrological model. *Hydrol. Sci. J.* **2019**, *64*, 1472–1487. [CrossRef]
76. Megnounif, A.; Terfous, A.; Ouillon, S. A graphical method to study suspended sediment dynamics during flood events in the Wadi Sebdou, NW Algeria (1973–2004). *J. Hydrol.* **2013**, *497*, 24–36. [CrossRef]
77. Laraque, A.; Olivry, J.C. Evolution de l'hydrologie Du Congo-Zaïre et de Ses Affluents Rive Droite et Dynamique Des Transports Solides et Dissous. *IAHS Publ.* **1996**, *238*, 271–288.
78. Laraque, A.; Maziezoltla, B. *Banque de Données Hydrologiques Des Affluents Congolais Du Fleuve Congo-Zaïre et Informations Physiographiques*; Programme PEGI/GBF, Volet Congo-UR22/DEC; ORSTOM—Laboratorie D'hydrologie: Montpellier, France, 1995; p. 250.
79. Aich, V.; Zimmermann, A.; Elsenbeer, H. Quantification and interpretation of suspended-sediment discharge hysteresis patterns: How much data do we need? *Catena* **2014**, *122*, 120–129. [CrossRef]
80. Malutta, S.; Kobiyama, M.; Chaffe, P.L.B.; Bonumá, N.B. Hysteresis analysis to quantify and qualify the sediment dynamics: State of the art. *Water Sci. Technol.* **2020**, *81*, 2471–2487. [CrossRef]
81. Nguimalet, C.-R.; Orange, D. Recent Hydrological Dynamics of Oubangui at Bangui (Central African Republic): Anthropogenic or Climatic Impacts? *Geo. Eco. Trop.* **2013**, *37*, 101–112.
82. Runge, J. The Congo River, Central Africa. *Large Rivers* **2008**, 293–309. [CrossRef]
83. Carr, A.B.; Trigg, M.A.; Tshimanga, R.M.; Borman, D.J.; Smith, M.W. Greater Water Surface Variability Revealed by New Congo River Field Data: Implications for Satellite Altimetry Measurements of Large Rivers. *Geophys. Res. Lett.* **2019**, *46*, 8093–8101. [CrossRef]
84. Marco, P.; Van Griensven, A. *Suitability of SWAT Model for Sediment Yields Modelling in the Eastern Africa*; Technical Paper; University of Dares Salam: Dares Salaam, Tanzania, 2011. [CrossRef]
85. Mayaux, P.; Bartholomé, E.; Massart, M.; Van Cutsem, C.; Cabral, A.; Nonguierma, A.; Diallo, O.; Pretorius, C.; Thompson, M.; Cherlet, M.; et al. A Land Cover Map of Africa. Carte de l'occupation Du Sol de l'Afrique. *EUR* **2003**, *20665*.
86. Ngongo, M.; Kasongo, E.; Muzinya, B.; Baert, G.; Van Ranst, E. Soil Resources in the Congo Basin: Their Properties and Constraints for Food Production. *Nutr. Food Prod. Congo Basin* **2014**, *c*, 214.
87. O'Loughlin, F.; Neal, J.; Schumann, G.; Beighley, E.; Bates, P. A LISFLOOD-FP hydraulic model of the middle reach of the Congo. *J. Hydrol.* **2020**, *580*, 124203. [CrossRef]
88. O'Loughlin, F.; Trigg, M.A.; Schumann, G.J.-P.; Bates, P.D. Hydraulic characterization of the middle reach of the Congo River. *Water Resour. Res.* **2013**, *49*, 5059–5070. [CrossRef]
89. Molliex, S.; Kettner, A.J.; Laurent, D.; Droz, L.; Marsset, T.; Laraque, A.; Rabineau, M.; N'Kaya, G.D.M. Simulating sediment supply from the Congo watershed over the last 155 ka. *Quat. Sci. Rev.* **2019**, *203*, 38–55. [CrossRef]
90. Garzanti, E.; Vermeesch, P.; Vezzoli, G.; Andò, S.; Botti, E.; Limonta, M.; Dinis, P.; Hahn, A.; Baudet, D.; De Grave, J.; et al. Congo River sand and the equatorial quartz factory. *Earth Sci. Rev.* **2019**, *197*, 102918. [CrossRef]
91. Laraque, A.; Pouyaud, B.; Rocchia, R.; Robin, R.; Chaffaut, I.; Moutsambote, J.M.; Maziezoula, B.; Censier, C.; Albouy, Y.; Elenga, H.; et al. Origin and function of a closed depression in equatorial humid zones: The lakeTele in north Congo. *J. Hydrol.* **1998**, *207*, 236–253. [CrossRef]
92. Haensler, A.; Jacob, D.; Kabat, P.; Ludwig, F. *Climate Change Scenarios for the Congo Basin*; Climate Service Centre Report No. 11; Climate Service Center: Hamburg, Germany, 2013; ISBN 2192-4058.
93. Filizola, N.; Guyot, J.L. Suspended sediment yields in the Amazon basin: An assessment using the Brazilian national data set. *Hydrol. Process.* **2009**, *23*, 3207–3215. [CrossRef]
94. Meade, R.H.; Dunne, T.; Richey, J.E.; Santos, U.D.M.; Salati, E. Storage and Remobilization of Suspended Sediment in the Lower Amazon River of Brazil. *Science* **1985**, *228*, 488–490. [CrossRef] [PubMed]
95. Richey, J.E.; Meade, R.H.; Salati, E.; Devol, A.H.; Nordin, C.F.; Dos Santos, U. Water Discharge and Suspended Sediment Concentrations in the Amazon River: 1982–1984. *Water Resour. Res.* **1986**, *22*, 756–764. [CrossRef]

96. Wei, X.; Sauvage, S.; Ouillon, S.; Le, T.P.Q.; Orange, D.; Herrmann, M.; Sanchez-Perez, J.-M. A modelling-based assessment of suspended sediment transport related to new damming in the Red River basin from 2000 to 2013. *Catena* **2021**, *197*, 104958. [CrossRef]
97. Hughes, D.A.; Tshimanga, R.M.; Tirivarombo, S.; Tanner, J. Simulating wetland impacts on stream flow in southern Africa using a monthly hydrological model. *Hydrol. Process.* **2013**, *28*, 1775–1786. [CrossRef]
98. Sun, X.; Bernard-Jannin, L.; Garneau, C.; Volk, M.; Arnold, J.G.; Srinivasan, R.; Sauvage, S.; Sánchez-Pérez, J.M. Improved simulation of river water and groundwater exchange in an alluvial plain using the SWAT model. *Hydrol. Process.* **2016**, *30*, 187–202. [CrossRef]
99. Sun, X.; Bernard-Jannin, L.; Grusson, Y.; Sauvage, S.; Arnold, J.; Srinivasan, R.; Pérez, J.M.S. Using SWAT-LUD Model to Estimate the Influence of Water Exchange and Shallow Aquifer Denitrification on Water and Nitrate Flux. *Water* **2018**, *10*, 528. [CrossRef]
100. Sok, T.; Oeurng, C.; Ich, I.; Sauvage, S.; Sánchez-Pérez, J.M. Assessment of Hydrology and Sediment Yield in the Mekong River Basin Using SWAT Model. *Water* **2020**, *12*, 3503. [CrossRef]

Review

Review of a Semi-Empirical Modelling Approach for Cohesive Sediment Transport in River Systems

Bommanna G. Krishnappan 

Krishnappan Environmental Consultancy, Hamilton, ON L9C 2L3, Canada; krishnappan@sympatico.ca

Abstract: In this paper, a review of a semi-empirical modelling approach for cohesive sediment transport in river systems is presented. The mathematical modelling of cohesive sediment transport is a challenge because of the number of governing parameters controlling the various transport processes involved in cohesive sediment, and hence a semi-empirical approach is a viable option. A semi-empirical model of cohesive sediment called the RIVFLOC model developed by Krishnappan is reviewed and the model parameters that need to be determined using a rotating circular flume are highlighted. The parameters that were determined using a rotating circular flume during the application of the RIVFLOC model to different river systems include the critical shear stress for erosion of the cohesive sediment, critical shear stress for deposition according to the definition of Partheniades, critical shear stress for deposition according to the definition of Krone, the cohesion parameter governing the flocculation of cohesive sediment and a set of empirical parameters that define the density of the floc in terms of the size of the flocs. An examination of the variability of these parameters shows the need for testing site-specific sediments using a rotating circular flume to achieve a reliable prediction of the RIVFLOC model. Application of the model to various river systems has highlighted the need for including the entrapment process in a cohesive sediment transport model.

Keywords: cohesive sediment transport; modelling; flocculation; critical shear stress; erosion; deposition; river systems; entrapment

Citation: Krishnappan, B.G. Review of a Semi-Empirical Modelling Approach for Cohesive Sediment Transport in River Systems. *Water* **2022**, *14*, 256. https://doi.org/10.3390/w14020256

Academic Editor: Achim A. Beylich

Received: 15 December 2021
Accepted: 14 January 2022
Published: 16 January 2022

Publisher's Note: MDPI stays neutral with regard to jurisdictional claims in published maps and institutional affiliations.

1. Introduction

Cohesive sediments play an important role in the transportation of contaminants and nutrients in a river system. Cohesive sediments are characterized as a mixture of predominantly clay- and silt-sized fractions of clay-type minerals but may also contain a range of organic compounds [1]. Contaminants and nutrients interact with these cohesive sediment mixtures and become part of the assemblage of the mineral and organic particles due to physical, chemical and biological controls [2]. Therefore, the contaminants and nutrients are transported in the river systems predominantly in association with cohesive sediments. A better understanding of the transport processes of cohesive sediment is of paramount importance for understanding the water quality and the quality of the river ecosystem.

Transport processes of cohesive sediment were studied extensively in the literature for over sixty years. The pioneering work in this area was carried out by Partheniades [3–5], Partheniades and Kennedy [6] and Partheniades et al. [7]. Partheniades used a rotating circular flume to study the physical behavior of cohesive sediments in the laboratory and concluded that the cohesive sediment behavior is very different from that of the cohesionless coarse-grained sediments, which had been studied extensively in this area of research (see, e.g., [8–10]). Among the differences in the transport behaviors between cohesive and cohesionless sediments, the one that sets them apart is the flocculation process. Because of the fineness of the cohesive sediments, the surface forces of attraction and repulsion become the same order of magnitude as the body (gravity) forces. As a consequence, the cohesive sediments undergo the process of flocculation in which the cohesive sediment particles interact and become agglomeration of particles called flocs [3,7,11–15]. Cohesionless sediments, on the other hand, behave as individual particles in a flow field.

Analysis of cohesive sediment transport in a flow field is much more difficult in comparison to the analysis of cohesionless sediment. Since the cohesionless sediment behaves as individual particles and the density of the particles is also constant (~2.65), the settling velocity of the cohesionless sediment is well defined for a sediment of a certain size (Stokes Law). The same cannot be said for the cohesive sediment. In the case of cohesive sediment, because of the flocculation process, the size of the transporting unit in a flow field is a variable and depends, among other things, on the intensity of the flow field, which can break up the flocs into smaller units. Moreover, the density of the transported unit is also variable as a result of the incorporation of suspending fluid into the floc structure. Therefore, the specification of settling velocity for cohesive sediment is much more difficult than for the cohesionless sediment.

In addition, the erosion and deposition characteristics of cohesive and cohesionless sediments are also different. From his laboratory studies, Partheniades [3] concluded that for cohesive sediments, the critical shear stress for erosion is different from the critical shear stress for deposition, and at a particular bed shear stress, cohesive sediments undergo either erosion or deposition but not both simultaneously. Cohesionless sediments, on the other hand, have only one critical condition that is valid for both erosion and deposition and undergo simultaneous erosion and deposition processes under all bed shear stress conditions.

Earlier research had treated the cohesive sediment as part of the "wash load", which was defined as the sediment load consisting of grain sizes that are considerably finer than those present in the stream bed [8]. It was further assumed that the wash load is "supply limited" and its transport rate is independent of flow characteristics. However, recent research has shown that the cohesive sediments do interact with stream beds and the transport characteristics of cohesive sediments, including the flocculation mechanism, do depend on flow characteristics [3,4,7,11–17].

When modelling the transport of cohesive sediments in rivers, the aforementioned differences between the cohesive and cohesionless sediments have to be taken into account rather than simply extending the cohesionless sediment transport models to cohesive sediment as has often been done in the literature. In cohesionless transport models, a mass balance equation is usually solved with the critical shear stress for erosion, as given by the Shield's diagram and employing any one of the numerous sediment transport formulae that have been derived experimentally for these types of sediments. For cohesive sediments on the other hand, the critical shear stress conditions for erosion and deposition that are universally accepted are not available and the sediment transport rate functions are also not readily available. The flocculation process also needs to be taken into account. Under these circumstances, the modelling of cohesive sediment transport in river systems requires a fresh approach that will address the unique nature of cohesive sediment transport processes such as flocculation and the distinctive erosion and deposition processes that the sediment experiences over a certain range of bed shear stresses prevailing in a river system. An examination of the flocculation process and the erosion deposition processes is undertaken here to highlight the complexities of a cohesive sediment transport model.

1.1. Flocculation Process

Early studies on the flocculation of cohesive sediments were carried out in estuarial systems where the mixing of freshwater from the river with the salt water in the ocean caused the river sediments to flocculate [3,18]. In these studies, the flocculation process was treated as a two-step process; in step one, called the "collision process", the particles are made to collide against each other by processes such as Brownian motion, turbulence, velocity gradients, inertia and differential settling, and in step two, called the "cohesion process" the collided particles are bonded together to form an agglomeration of particles called flocs. In estuarine systems, the bonding or the cohesion was provided by the relative strengths of attractive versus repulsive forces between the particles. The attractive forces are due to van der Waal's forces, which vary inversely as the seventh power of the distance between the particles. The repulsive forces are due to the like charges of the ion clouds surrounding the clay particles [18].

In recent studies on flocculation, cohesive sediments were observed to form flocs even in freshwater systems [19–21]. The cohesion mechanism involved in the freshwater flocculation is different from the one found in estuarine systems. In freshwater flocculation, bacteria and other microorganisms were found to play a role in the floc formation [2,21,22]. The microorganisms secrete polymers [23], which provide the bonding among particles. The mechanism of flocculation by polymers can be explained on the basis of the classical interparticle bridging model described in Ruehrwein and Ward [24] and La Mer and Healy [25]. Busch and Stumm [26] reported that the amount of polymers required for optimal flocculation is extremely small, of the order of a mg/L.

1.2. Distinct Nature of Erosion and Deposition Processes of Cohesive Sediment

A better understanding of the erosion and deposition processes governing the transport of sediment in a river system is important for modelling the sediment fluxes at the sediment–water interface. The settling and the dispersive fluxes at the sediment–water interface is balanced by the net amount of sediment entering the flow domain from the bed. For cohesionless sediments at a given bed shear stress greater than the critical shear stress for erosion, the deposition and erosion of the sediment occur simultaneously, and for a steady state condition, the deposition flux is equal to the erosion flux. For the cohesive sediment, on the other hand, since the deposition and the erosion processes are mutually exclusive, there is only one flux at a time, either a deposition flux or an erosion flux. The existence of two different critical conditions for the cohesive sediment can be explained as follows: when a cohesive sediment floc deposits onto a bed consisting of cohesive sediment flocs, the deposited floc sticks to the flocs that are part of the bed due to cohesion and requires a slightly higher bed shear stress to dislodge the deposited floc from the bed. Therefore, for cohesive sediments, two distinctive critical shear stresses can be identified, one for deposition and the other for erosion: the critical shear stress for erosion is always higher than the critical shear stress for deposition.

The erosion characteristics of cohesive sediment have been studied extensively in the literature [3,6,7,11–14,27–31]. These studies have shown that the erosion characteristics of cohesive sediment deposits depend on a number of parameters, including the manner in which the sediment deposit is created, the time of consolidation, the rate of application of the bed shear stress and the stabilizing effects of the microorganisms. A comprehensive list of all the governing parameters that control the erosion process of cohesive sediment deposits can be found in Hayter [32].

Because of the numerous controlling parameters involved in both the flocculation process and the erosion and deposition processes of cohesive sediment, it is practically impossible to derive analytical expressions representing the flocculation and the erosion and deposition processes of cohesive sediment transport. Therefore, the approach of a semi-empirical formulation of a mathematical model to represent the transport of cohesive sediment becomes attractive for finding practical solutions to problems pertaining to the transport of cohesive sediment and the associated contaminants. Krishnappan [33,34] developed a cohesive sediment transport model for rivers (RIVFLOC) and employed a semi-empirical approach to determine the parameters of the model by carrying out experiments in a rotating circular flume for site-specific sediment and the river water. Using these parameters, the model can then be applied to the actual rivers to predict the transport of cohesive sediment and the associated contaminants. This modelling approach had been used successfully to a number of river systems [34–39]. In this review paper, the RIVFLOC model of Krishnappan [33,34] is described highlighting the model parameters and their determination using a rotating circular flume.

The variability of the model parameters of different river systems is also examined. The objective of this review is to provide guidance for modelling cohesive sediment transport and the associated contaminants in a river system.

2. Materials and Methods

2.1. Description of the RIVFLOC Model

The RIVFLOC model predicts the transport of fine-grained, cohesive sediments in rivers. It has two components: a transport and dispersion component and a flocculation component. The transport and dispersion component is based on the advection-diffusion equation expressed in a curvilinear co-ordinate system to treat the transport and mixing of fine sediment entering a river as steady source. The flocculation component adapts a coagulation equation that includes four different collision mechanisms and treats the process of agglomeration of the fine sediment as it is transported in a river flow. Details of the model are described in Krishnappan [33]. Here, some salient features of the model are described for the sake of completeness.

2.1.1. Transport and Dispersion Component

A depth-averaged advection-dispersion equation expressed in the curvilinear co-ordinate system (Figure 1) introduced by Yotsukura and Sayre [40] is used to analyze the transport and dispersion process of fine sediment in the RIVFLOC model. The form of the equation solved is shown below:

$$\frac{\partial C_k}{\partial x} = \frac{\partial}{\partial \eta}\left(\frac{M_x U h^2 E_z}{Q^2}\frac{\partial C_k}{\partial \eta}\right) + \frac{M_x \lambda_1}{U}C_k + \frac{M_x}{U}\lambda_2 \tag{1}$$

where

C_k = depth-averaged volumetric concentration of the kth size fraction.
x = distance along the longitudinal co-ordinate axis (Figure 1).
U = depth-averaged velocity component in the x direction.
h = depth of flow.
Q = flow rate.
E_z = transverse dispersion coefficient.
M_x, M_z = metric coefficients of the curvilinear co-ordinate system.
η = normalized cumulative discharge given by:
$\eta = \frac{1}{Q}\int_0^z M_z U h dz.$
z = transverse co-ordinate.
λ_1 = rate coefficient for reaction.
λ_2 = rate of sediment source or sink.

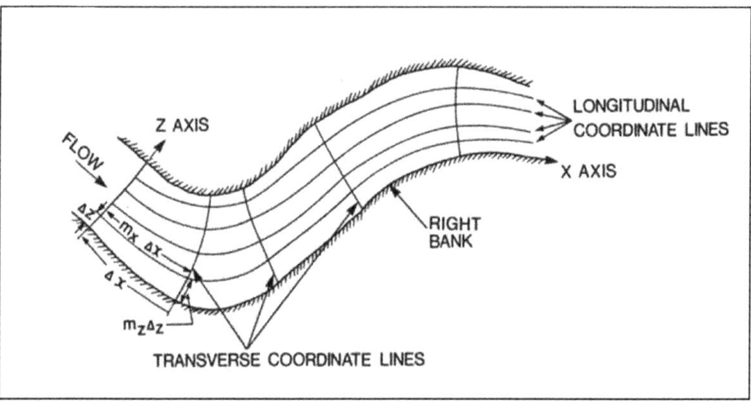

Figure 1. The curvilinear co-ordinate system used in RIVFLOC model.

Boundary conditions needed to solve the mass balance equation are as follows:

1. For the upstream boundary, a known transverse distribution of fine sediment entering the river has to be specified together with the size distribution of the sediment. This can be obtained either by direct measurement or from some other calculations.
2. At the side banks, the sediment fluxes crossing these banks are zero, and hence the concentration gradients across those boundaries are taken as zero.
3. At the bed, the sediment fluxes across the sediment–water interface need to be specified. Transfer of sediment between the bed and water column can occur because of the following three processes, namely, erosion, deposition and entrapment (ingress of fines in coarse bed sediment). The first two processes were studied extensively in the literature. For example, the erosion rate of fine sediment has been studied by Partheniades [3], Mehta and Partheniades [41], Parchure [30], Lick [13], and Krishnappan et al. [42]. Deposition characteristics of fine sediment were studied by Krone [11], Mehta and Partheniades [12], Lick [13], among others.

In the RIVFLOC model, Krone's approach was used to specify the net deposition of fine sediment to the river bed. According to this approach, the net deposition flux (q_d) was calculated as:

$$q_d = Pw_s C_{kb} \tag{2}$$

where P represents the probability that a floc, settling to the bed, stays at the bed (a stickiness parameter). w_s is the settling velocity of sediment and C_{kb} is the near-bed concentration of the kth size fraction of the sediment. Krone [11] proposed a relationship for P as:

$$P = \left(1 - \frac{\tau}{\tau_{crdk}}\right) \text{ for } \tau < \tau_{crdk} \text{ and } P = 0 \text{ for } \tau > \tau_{crdk} \tag{3}$$

where τ is the bed shear stress and τ_{crdk} is the critical shear stress for deposition, which was defined by Krone as the bed shear stress above which none of the initially suspended sediment would deposit. The bed shear stress τ (together with the flow field) has to be determined using a hydrodynamic model for the river flow. The critical shear stress for deposition has to be determined experimentally for site-specific sediments using specialized experimental facilities such as a rotating circular flume.

For bed shear stresses greater than τ_{crdk}, deposition flux will be zero, but erosion flux can exist. In the RIVFLOC model, the erosion flux was modelled using an erosion rate proposed in Krishnappan et al. [42]. Accordingly,

$$q_e = E \tag{4}$$

where E is given by:

$$E = h \frac{(1/c_1)}{\left(\frac{c_0}{c_1}t + 1\right)^2} \tag{5}$$

where c_0 and c_1 are constants to be determined from laboratory experiments using a rotating flume for site-specific sediment and t is the elapsed time from the stress change during a particular shear stress step.

Unlike erosion and deposition, the entrapment process is not studied extensively in the literature. In an empirical study carried out by Krishnappan and Engel [43], it was shown that the entrapment flux can be related to the settling flux through a proportionality coefficient, termed entrapment coefficient. According to this study, the entrapment flux (q_{entrap}) can be is expressed as:

$$q_{entrap} = \alpha w_s C_{kb} \tag{6}$$

where α is the entrapment coefficient. This entrapment coefficient can be a function of porosity of the gravel substrate, thickness of the gravel bed layer and the permeability of the gravel substrate, but a quantitative relationship involving these parameters is not available at the present time.

Equation (1) was solved using a finite difference methodology developed by Stone and Brian [44]. The accuracy of the scheme was analyzed by Krishnappan and Lau [45].

2.1.2. Flocculation Stage

The flocculation stage was based on a coagulation equation (Fuchs [46]) shown below:

$$\frac{\partial N(i,j)}{\partial t} = -\beta N(i,t)\sum_{j=1}^{\infty}K(i,j)N(j,t) + \frac{1}{2}\beta\sum_{j=1}^{\infty}K(i-j,j)N(i-j,j)N(j,t) \tag{7}$$

where $N(i,j)$ and $N(j,t)$ are number concentrations of particle size classes i and j, respectively. $K(i,j)$ is the collision frequency function, which is a measure of the probability that a particle of size i collides with a particle of size j in unit time, and β is the cohesion factor, which defines the probability that a pair of collided particles will coalesce and form a new floc. The cohesion factor accounts for the different cohesion mechanisms such as the electrochemical cohesion that occurs in estuaries where the freshwater meets the salt water and in the freshwater systems, where the cohesion occurs due to polymers secreted by micro-organisms, etc. In this study, β is treated as an empirical parameter and was determined through the calibration of the model using experimental data.

Equation (7) was simplified by considering the particle sizes in discrete size classes as shown in Figure 2. In this figure, the continuous particle size space is considered in discrete size ranges (1 to M). Each range is considered as a bin containing particles of a certain size range. For example, r_1 is the geometric mean radius of particles in bin 1. The particle ranges were selected in such a way that the mean volume (v_i) of particles in bin i is twice that of the preceding bin. The volume ranges of the various size ranges are shown in Figure 2.

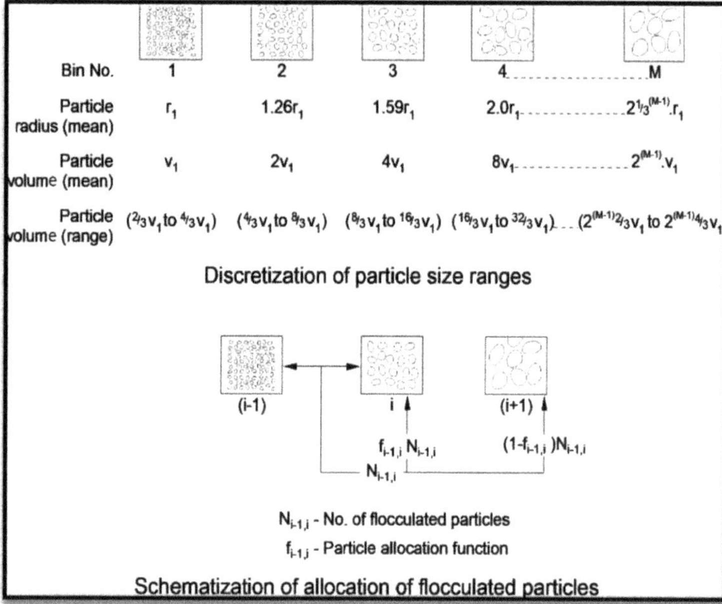

Figure 2. Schematic representation of flocculation stage.

Under this scheme, when particles of bin i flocculate with particles in bin j ($j < i$), the newly formed particles will fit into bins i and $i + 1$. The proportion of particles belonging to these bins can be calculated by considering the mass balance of particles before and after flocculation (allocation function). With this simplification, the coagulation equation is expressed in discrete form as follows:

$$\frac{\Delta N_i}{\Delta t} = -\sum_{j \neq i} \beta K(r_i, r_j) N_i N_j + \sum_{j < i} f_{i,j} \beta K(r_i, r_j) N_i N_j + \sum_{j < i-1} (1 - f_{i,j}) \beta K(r_i, r_j) N_{i-1} N_j \quad (8)$$

where $f_{i,j}$ is the allocation function given by:

$$f_{i,j} = \left(\rho_{s,i} V_i + \rho_{s,j} V_j - \rho_{s,j+1} V_{i+1} \right) / \left(\rho_{s,j} V_i - \rho_{j+1} V_{i+1} \right) \quad (9)$$

where ρ_s is the density of the flocs and V is the volume of the flocs. The density of the flocs is dependent on their size, and an empirical relationship between the two was formulated by Lau and Krishnappan [47]. This relationship was adopted for the present study. The form of the relationship is as follows:

$$\rho_s - \rho = \left(\rho_p - \rho \right) \exp(-bD^c) \quad (10)$$

where ρ_s, ρ and ρ_p are densities of flocs, water and parent material forming the flocs, respectively. D is the diameter of the floc and b and c are empirical coefficients to be determined from rotating flume experiments in the laboratory. The settling velocity of flocs, needed for the settling stage of the particle motion, was calculated using Equation (10) and the Stoke's Law. The resulting expression is as follows:

$$w_k = (1.65/18) \left(g D_k^2 / v \right) \exp(-b D_k^c) \quad (11)$$

where g represents the acceleration due to gravity and ϑ stands for the kinematic viscosity of water.

The collision frequency function $K(i, j)$ assumes different functional forms depending on the collision mechanisms that bring the particles to close proximity. The mechanisms considered in this study are (1) Brownian motion (K_b), (2) turbulent fluid shear (K_{sh}), (3) inertia of particles in turbulent flow (K_I) and (4) differential settling of flocs (K_{ds}). An effective collision frequency function, K_{ef}, was calculated in terms of the individual collision functions as follows:

$$K_{ef} = K_b + \sqrt{\left(K_{sh}^2 + K_I^2 + K_{ds}^2 \right)} \quad (12)$$

The individual collision frequency functions were formulated by Valioulis and List [48], and the form of the functions are as follows:

For Brownian motion:

$$K_b(r_i, r_j) = \frac{2}{3} \frac{KT}{\mu} \frac{(r_i - r_j)^2}{r_i r_j} \quad (13)$$

For fluid shear:

$$K_{sh}(r_i, r_j) = \frac{4}{3} \left(\frac{\epsilon}{v} \right)^{1/2} (r_i - r_j)^3 \quad (14)$$

For inertia:

$$K_I(r_i, r_j) = 1.21 \frac{\rho_s}{\rho_f} \left(\frac{\epsilon^3}{v^5} \right)^{1/4} (r_i + r_j)^2 \left| r_i^2 - r_j^2 \right| \quad (15)$$

For differential settling:

$$K_{ds}(r_i, r_j) = \frac{2}{9} \frac{\pi g}{v} \left(\frac{\rho_s - \rho_f}{\rho_f} \right) (r_i + r_j)^2 \left| r_i^2 - r_j^2 \right| \quad (16)$$

The symbols appearing in the above collision frequency functions have the following meanings: K is the Boltzmann constant, T is the absolute temperature in degree kelvin, μ is the viscosity of fluid, v is the kinematic viscosity, ϵ is the turbulent energy dissipation per unit mass, and ρ_s and ρ_f are densities of sediment and fluid, respectively.

The break-up of flocs due to turbulent fluctuations of the flow was modelled using a method proposed by Tambo and Watanabe [49]. According to this method, a "collision-agglomeration" function was introduced as a multiplier to the collision frequency function, K_{ef}. This resulted in an optimum floc size distribution for a given level of turbulence. The function proposed by Tambo and Watanabe is given below:

$$\alpha_R = \alpha_0 \left(1 - \frac{R}{(R_m + 1)}\right)^n \tag{17}$$

where R is the number of primary particles in a floc and R_m is the number of particles in the biggest floc for the given level of turbulence. The parameters α_0 and n take on values of 1/3 and 6, respectively.

The input data requirement for the RIVFLOC model and the possible sources of such data are summarized in Table 1 below:

Table 1. Input data requirement for the RIVFLOC model.

Number	Input Data Needed for RIVFLOC Model	Possible Sources of Such Data
1	The plan form of river- reach to supply the metric coefficient M_x	A field survey to measure the cross sections at a number of locations along the river reach.
2	Flow characteristics such as h, U and Q	A field measurement of h, U and Q or use the computational method described in Section 2.2.
3	Upstream boundary condition for sediment concentration	A field measurement at the upstream boundary of the model domain.
4	Upstream boundary condition for the size distribution of the sediment	A field measurement of size distribution of suspended sediment using instruments such as LISST (see Stone et al. [38]).
5	Transverse dispersion coefficient E_z	Literature data on transverse mixing in rivers.
6	Critical shear stress for erosion and deposition	Laboratory measurements using a rotating circular flume described in Section 2.3.
7	Cohesion parameter, β	Laboratory measurements using a rotating circular flume described in Section 2.3.
8	Empirical parameters b and c appearing in the relationship floc size vs. floc density	Laboratory measurement using a rotating circular flume described in Section 2.3.

2.2. A Computational Methodology for the Lateral Variation of the Longitudinal Velocity U

The development of the curvilinear coordinate system shown in Figure 1 requires knowledge of the depth-averaged longitudinal velocity component (U) distribution in the lateral direction and the depth variation across the river. The depth variation across the river can be obtained from the cross-sectional surveys. The depth-averaged velocity distributions are computed within the RIVFLOC model using a methodology proposed by Djordjevic [50] and Guan et al. [51]. According to this method, the depth-averaged momentum equation in the longitudinal direction is simplified as follows:

$$\frac{\partial(hU^2)}{\partial x} + \frac{\partial(hUV)}{\partial y} = -gh\frac{dz_w}{dx} + \frac{1}{\rho}\frac{\partial(h\tau_{xy})}{\partial y} - \frac{1}{\rho}\tau_{bx} \tag{18}$$

In the equation above, V is the depth-averaged velocity components in the transverse directions, z_w is the water surface elevation above a datum, g and ρ are the gravitational acceleration and water density, respectively, τ_{xy} is the depth-averaged turbulent shear stress and τ_{bx} is the bed shear stress. The depth-averaged turbulent shear stress is calculated using the eddy viscosity concept as:

$$\frac{\tau_{xy}}{\rho} = \vartheta_t \frac{\partial U}{\partial y} \tag{19}$$

where ϑ_t is the depth-averaged eddy viscosity coefficient. Assuming similarity with mass transport, ϑ_t is expressed as:

$$\vartheta_t = c_v h U_* \tag{20}$$

where c_v is an empirical constant and U_* is the shear velocity. The bed shear stress is expressed as:

$$\frac{\tau_{bx}}{\rho} = \frac{\tau_b}{\rho \cos \varnothing} = \frac{U_*^2}{\cos \varnothing} = c_f \frac{U_*^2}{\cos \varnothing} \tag{21}$$

where $\varnothing$ is the transverse angle of inclination of the bed to the horizontal, c_f is the friction factor, which can be expressed in terms of the Manning's n as:

$$c_f = g n^2 R^{1/3} = g n^2 \left(\frac{h \, dy}{dy / \cos \varnothing} \right)^{1/3} = g n^2 (h \cos \varnothing)^{1/3} \tag{22}$$

Substituting Equations (19)–(21) into Equation (18), we get:

$$g h \frac{dz_w}{dx} - \frac{c_v}{2} \frac{\partial}{\partial y} \left(h^2 \sqrt{c_f} \frac{\partial U^2}{\partial y} \right) + c_f \frac{U^2}{\cos \varnothing} = -\frac{\partial (h U^2)}{\partial x} - \frac{\partial (h U V)}{\partial y} \tag{23}$$

Assuming that the first term on the right-hand side of Equation (23) is small in comparison to the second term, and using Guan et al. [51] approximation for the product UV in the second term as:

$$UV = c_{uv} U^2 \tag{24}$$

Equation (23) can be expressed as follows:

$$\frac{d}{dy} \left(\frac{c_v}{2} n h^{11/6} (\cos \varnothing)^{1/6} \sqrt{g} \frac{dU^2}{dy} \right) - \frac{\partial}{\partial y} \left(c_{uv} h U^2 \right) - g n^2 h^{\frac{1}{3}} (\cos \varnothing)^{\frac{4}{3}} U^2 = g h \frac{dz_w}{dx} \tag{25}$$

The above equation is a second order ordinary differential equation with variable coefficients and can be solved numerically for the longitudinal velocity component U once the water surface slope appearing on the right-hand side of the equation is specified. For the RIVFLOC model, the water surface slope in the river reach was supplied by a one-dimensional mobile boundary flow model MOBED developed by Krishnappan [52–54]. The MOBED model solves the full St. Venant's equations and a sediment mass balance equation to treat the coarse-grained sediment constituting the riverbed and calculates the bed level changes in addition to water level changes as a function of time and distance along the river reach. The model also supplies the bed shear stresses needed for the RIVFLOC model to model the cohesive sediment transport.

2.3. Determination of Model Parameters Using a Rotating Circular Flume

The model parameters that needed to be determined from laboratory experiments are: critical shear stresses for erosion and deposition, erosion rate, cohesion parameter, and the empirical coefficients defining the floc size and floc density relationship (see Table 1). For performing laboratory experiments with cohesive sediments, the rotating circular flumes are preferable than the straight flumes, as the pumping and other flow regulation devices that are associated with the straight flume are likely to disrupt the flocs and cause floc breakage. To avoid this from happening, rotating circular flumes are invariably used for cohesive sediment transport experiments in which the flocculation mechanism is often the main focus of the study.

As part of a cohesive sediment transport research programme at the National Water Research Institute (NWRI) in Burlington, Ontario, Canada, a 5.0 m diameter circular flume was installed. Complete details of the flume can be found in Krishnappan [55]. Here, some of the salient features of the flume are highlighted. The flume is 5.0 m in diameter, 30 cm wide and 30 cm deep. It rests on a rotating platform, which is 7.0 m in diameter. A counter rotating top cover fits inside the flume with close tolerance and makes contact with the water surface. The simultaneous rotation of the flume and the top cover in opposite directions produces a flow field that is nearly two dimensional. A photograph showing the top view of the flume is shown in Figure 3.

Figure 3. A photograph of the top view of the rotating circular flume.

The flow fields generated in this flume were studied extensively [56–58]. These studies have shown that the flows generated in the flume are nearly two dimensional and the bed shear stress distributions across the flume are fairly uniform, as shown in Figure 4.

In Figure 4, the bed shear stresses are plotted as a function of distance across the flume with rotational speed as the parameter. The points are measured values and the lines are calculated distributions using a 3D hydrodynamic flow model (PHOENICS) developed by Roston and Spalding [59]. It can be seen from this figure that the PHOENICS model predictions agree well with the measured values.

The flume is fitted with a Malvern Particle Size Analyser [60] for measuring the size distribution of suspended sediment flocs. The instrument was placed directly underneath the flume, and the short tube connects the sampling point to the sample cell of the instrument. The sample is drawn through the sample cell continuously by gravity, and the sample after passing through the sample cell flows into a reservoir from where it was pumped back into the flume. The arrangement of the instrument is shown in Figure 5 and a photograph of the set-up is shown in Figure 6.

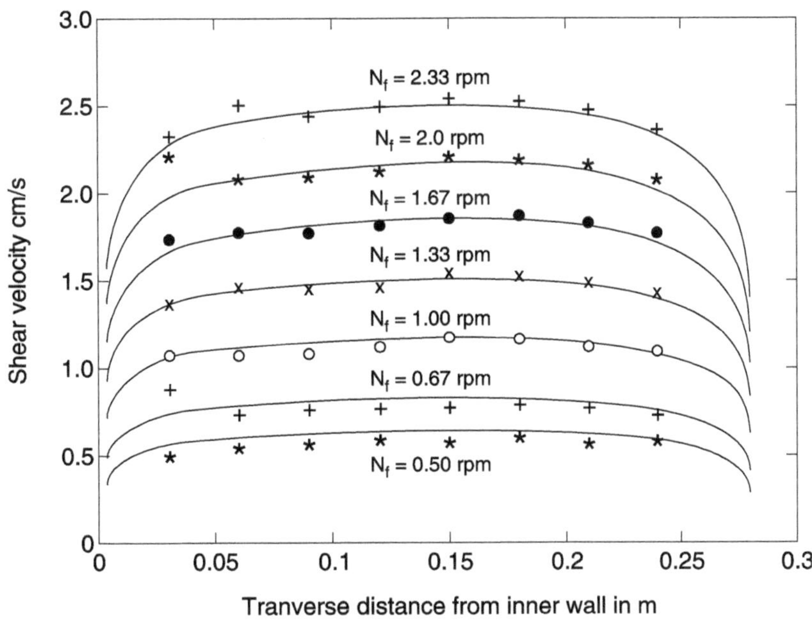

Figure 4. Bed shear stress distribution as a function of distance along the flume width.

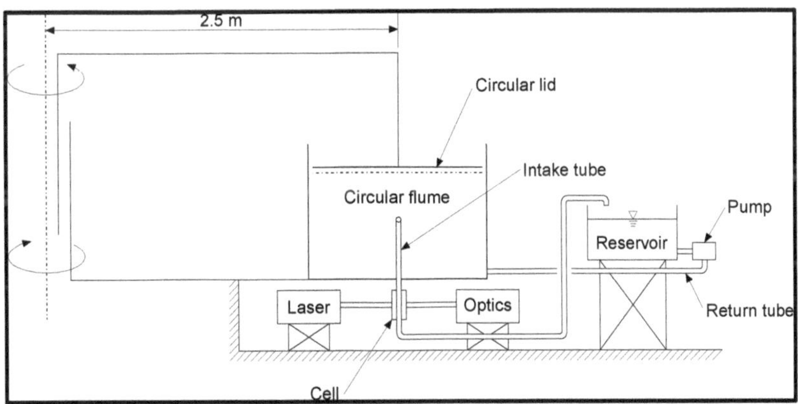

Figure 5. Schematic diagram showing the arrangement of the particle size analyzer mounted on the flume.

2.4. Collection of Sediment Samples for Testing in the Rotating Circular Flume

The samples of sediment–water mixtures were collected from the study areas in the field and were brought to the laboratory for testing in the rotating circular flume. The sampling device that was used for this purpose is described here. It is an inverted cone sampler specially designed for this purpose. The sampler consists of a conical chamber fitted with a propeller and a submerged pump. A photograph of the sampler is shown in Figure 7.

Figure 6. A photograph showing the placement of the Malvern instrument underneath the flume.

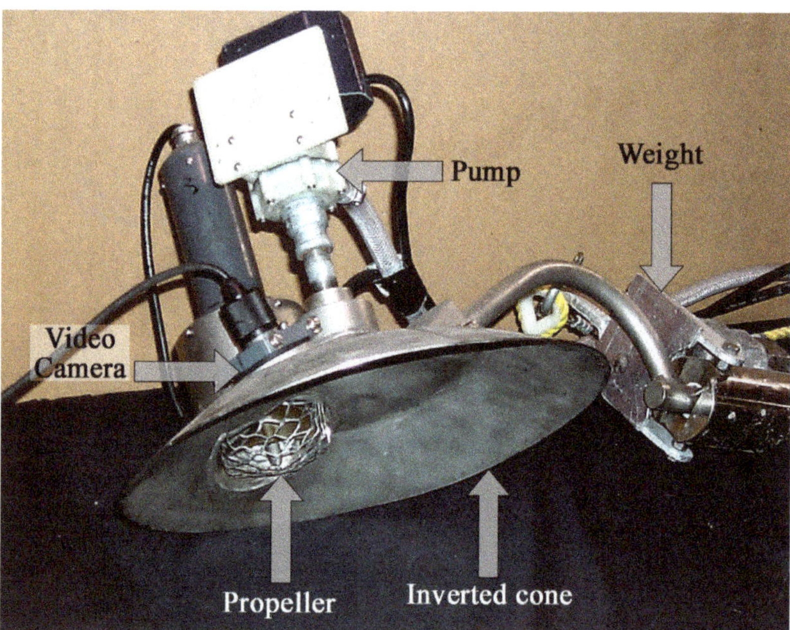

Figure 7. Inverted cone sampler for collecting samples of sediment–water mixture for testing in rotating circular flume.

To collect the sediment–water mixture using the sampler, the sampler was deployed in a depositional area in a river and the propeller and the pump were turned on remotely from a boat or from the shore. The propeller creates a sufficiently strong currents to dislodge the deposited fine sediment from the bed into suspension and the sediment–water mixture is pumped from the sampler to a sample container by operating the submerged pump. The sampler is also fitted with a weight to stabilize the sampler on the streambed against the flow and an under-water video camera to observe the operation of the sampler. Normally about five hundred to one thousand litres of sediment–water mixtures would be collected and transported to the laboratory in refrigerated trucks to preserve the chemical and biological characteristics of the sample.

2.5. Testing of Cohesive Sediments Transport Characteristics in the Rotating Circular Flume

Both depositional and erosional characteristics of cohesive sediment samples were tested in the rotating circular flume. For depositional experiments, sediment–water mixture was placed in the flume and the sediment concentration was adjusted by adding additional cohesive sediment to produce a fully mixed concentration of known value. Full mixing of the sediment was achieved by operating the flume at high speeds (2.5 rpm for the lid and 2 rpm for the flume, which corresponds to a bed shear stress of 0.6 Pa) that were found to be sufficient to maintain all of the sediment in suspension. The high-speed operation of the flume was maintained for a period of about 20 min and then the speeds were lowered to their respective test speeds to yield a particular bed shear stress in the flume. The flume was operated at this level for about 5 h. During this time, suspended sediment concentration and the size distribution of the suspended sediments were monitored at regular intervals of time. Experiments were then repeated for different bed shear stress conditions maintaining the initial concentration to be the same for all these tests. The influence of the initial concentration was also tested by conducting set-up experiments in which the initial concentration was varied while the bed shear stress was kept constant.

Erosion experiments were carried out by allowing all the suspended sediment to settle to the bed and allowing the deposited sediment to consolidate for different time durations. To begin an erosion test, the flume and the lid were started from rest and their speeds were increased in steps and each step was maintained for a period of 40 to 90 min. During each step, sediment samples were collected and the concentration of the eroded sediment in the water column were determined as a function of time. When the concentration of the sediment is sufficiently high, size distribution of the eroded sediment was also measured.

3. Results

The semi-empirical model of cohesive sediment transport (RIVFLOC model) was applied to different river systems to assess the cohesive sediment and associated contaminants in these systems. The listings of some of these studies are given in Table 2 along with the model parameters that were determined using the rotating flume for site-specific sediments.

Table 2. Listing of studies that employed the semi-empirical cohesive sediment transport model, RIVFLOC.

River Systems	τ_{cre} Pa	τ_{crdp} Pa	τ_{crdk} Pa	β	Empirical Constants		References
					b [1]	c [1]	
Hay River in NWT, Canada	0.14	0.08	0.40	0.010	0.02	1.35	Krishnappan [34]
Kingston, storm water pond, Canada	0.10	0.05	0.50	0.075	0.02	1.45	Krishnappan and Marsalek [35]
Ells River in Alberta, Canada	0.01	0.004	0.25	0.006–0.03	0.02	1.15	Droppo and Krishnappan [36]
Steepbank River in Alberta, Canada	0.13	0.000	0.30	0.000	0.61	0.35	Krishnappan [37]
Crowsnest River in Alberta, Canada	0.18	0.09	0.50	0.075	0.02	1.30	Stone et al. [38]
Taw River in UK	0.09	0.06	0.40	0.13–0.25	0.02–0.025	1.15–1.35	Stone et al. [39]

[1] Empirical constants appearing in Equation (10).

The methodologies used to determine the model parameters using the results from the depositional and erosional experiments for site-specific sediments carried out in the rotating circular flume are reviewed here in the following subsection. The experimental results for the Hay River sediment [34] are used as an example.

3.1. Determination of Critical Shear Stress for Deposition

The results from the deposition experiments carried out in the rotating circular flume for the Hay River sediment are summarized in Figure 8.

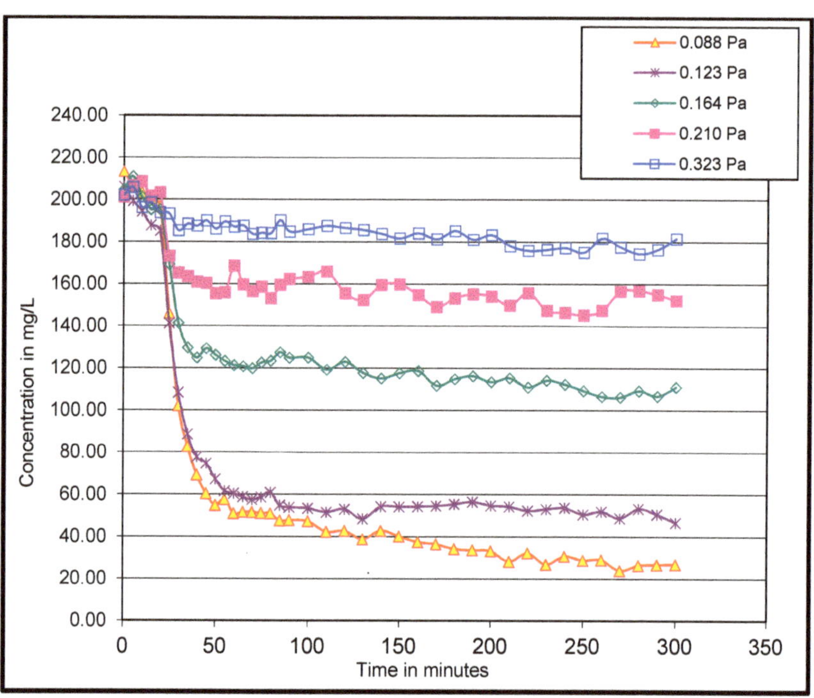

Figure 8. Depositional characteristics of the Hay River sediment from the North-West Territories in Canada.

Figure 8 shows the concentration variation of the suspended sediment as a function of time for different bed shear stresses as the sediment undergoes deposition. For a particular shear stress, concentration of sediment in suspension decreases initially as a function of time and reaches a steady state concentration, which is a function of the shear stress. When the shear stress is high, the steady state concentration is high, and a majority of initially suspended sediment stays in suspension (about 90% for the shear stress of 0.323 Pa). For lower shear stress tests, the concentration of sediment that stays in suspension diminishes, and for the lowest shear stress tested (0.088 Pa), the concentration of sediment in suspension is only about 10% of the initial sediment concentration. If the bed shear stresses were slightly lower than the lowest shear stress tested, all of the initially suspended sediment would have settled to the bed. Such a bed shear stress is termed the critical shear stress for deposition (τ_{crdp}: Partheniades' definition of critical shear stress for deposition). For Hay River sediment tested here, it can be determined by extrapolation that the critical shear stress for deposition was 0.08 Pa.

3.2. Determination of Critical Shear Stress for Erosion and the Erosion Rate

Results from the erosion experiments carried out for the Hay River sediments are shown in Figure 9.

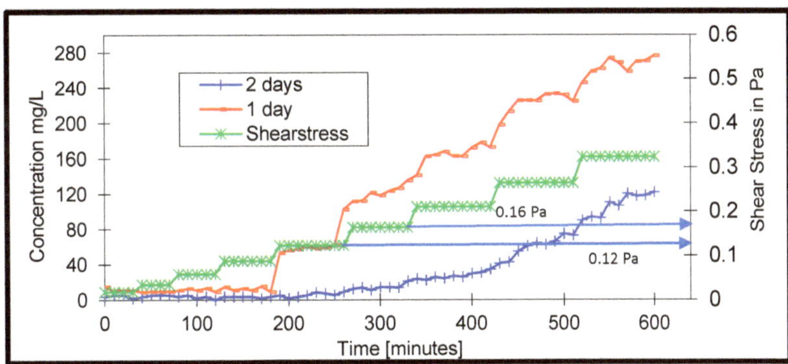

Figure 9. Results from the erosion experiments for the Hay River sediment.

The erosion experiments for the Hay River sediment were carried out for two different consolidation periods (one-day consolidation and two-day consolidation). As can be seen from Figure 9, for one-day consolidation, the onset of erosion occurs at a bed shear stress of 0.12 Pa, whereas for the two-day consolidation it is at 0.16 Pa, which shows that the consolidation process plays an important role in determining the erosion characteristics of Hay River sediments. For this sediment, the critical shear stress for erosion was taken as 0.14 Pa, which is an average of the two consolidation period experiments. The erosion rate function (i.e., Equation (5)) can be established for this sediment using the approach of Krishnappan et al. [42].

3.3. Influence of Initial Concentration on the Deposition Process

Experiments were carried in the rotating circular flume to test the influence of the initial concentration on the deposition process by varying the initial concentration and keeping the bed shear stress constant. Results from such an experiment is shown in Figure 10.

In Figure 10, the concentration variation as a function of time for two deposition experiments with the same bed shear stress but different initial concentrations (200 mg/L and 350 mg/L) are shown. From this figure, it is evident that the steady state concentration is a function of initial concentration. Higher initial concentration has resulted in higher steady state concentration. Such a behavior is typical of cohesive sediments because, for cohesionless sediments, the steady state concentration is independent of the initial concentration. It depends only on the bed shear stress. An explanation for such a behavior of cohesive sediment was offered by Partheniades and Kennedy [6]. They explained that during the deposition process of cohesive sediment, only flocs that are strong enough to withstand the high shear stress near the bed can deposit to the bed. Weaker flocs are susceptible for breakage under high shear and are likely to be broken into smaller flocs and brought back into suspension. In a sample of cohesive sediment, a certain fraction of the material can be made up of weaker flocs; hence, the higher the initial concentration, the higher the number of weaker flocs that can produce a higher steady state concentration. For this explanation to be valid, the concentration data in Figure 10, if normalized using the initial concentrations, should collapse into a single curve. Indeed, when the concentrations were normalized in terms of the initial concentrations, the two curves nearly collapsed into a single curve, as can be seen in Figure 11.

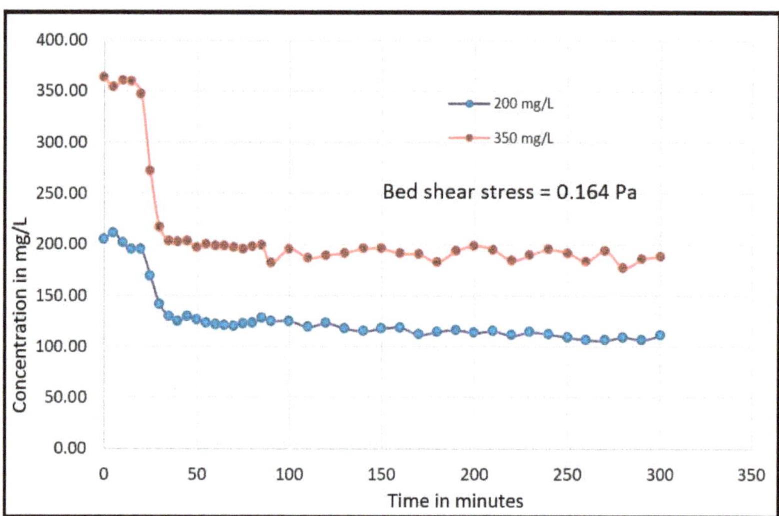

Figure 10. Influence of initial concentration on the deposition process.

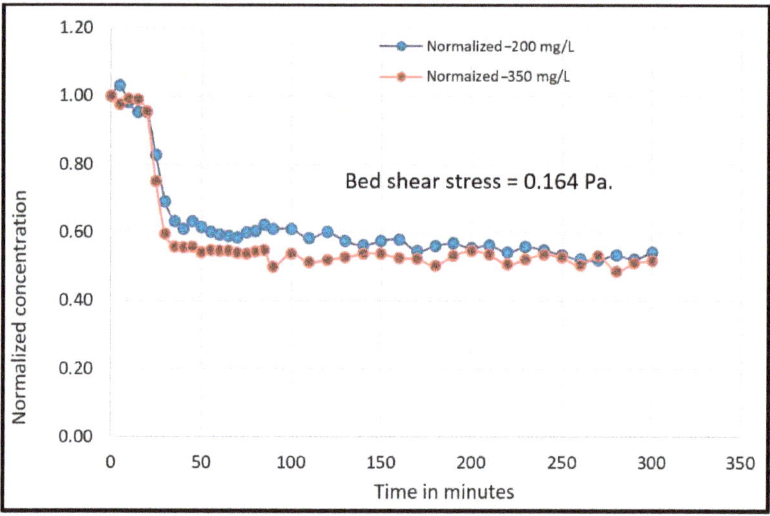

Figure 11. Normalized concentration during deposition.

When dealing with the deposition of cohesive sediments, it would be advantageous to treat the concentrations in terms of the initial concentrations, in which case the initial concentration ceases to be a governing parameter and the normalized concentration is a function of bed shear stress only as in the case of cohesionless sediment.

3.4. Size Distribution of Suspended Sediment Flocs

Size distribution data measured using the Malvern Particle Size Analyzer are summarized in Figure 12.

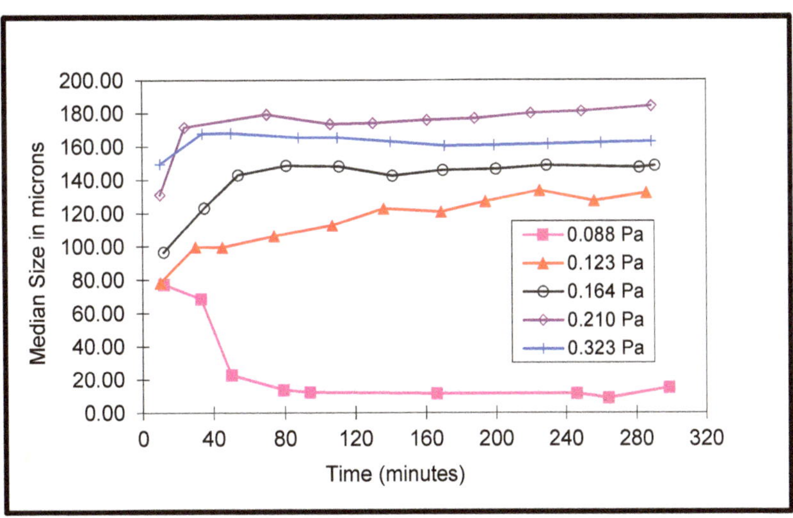

Figure 12. Median size of flocs as a function of bed shear stress.

In Figure 12, the median sizes of the distributions are plotted as a function of time for different bed shear stresses. At the lowest shear stress (i.e., at 0.088 Pa), the median size decreases as a function of time. This means that the distributions are becoming finer and finer as the time progresses. This implies that larger flocs with higher settling velocities settle out, leaving the smaller flocs with lower settling velocities in suspension. Such a behavior is characteristics of particles settling as individual particles without interaction among particles. As the bed shear stress increases, a very different picture emerges. For example, for the bed shear stress of 0.123 Pa, the median size of the distributions shows an increasing trend with time, suggesting that the flocculation of sediment is taking place. For this and higher bed shear stresses, the median size increases initially with time and then it levels off towards a steady state value. The increase in the steady state median size of flocs does not continue monotonically as a function of the bed shear stress. It reaches a maximum value at a shear stress of 0.210 Pa. With a further increase in shear stress, the median size actually decreases. This is due to the fact that, at higher shear stresses, the flocs are breaking up due to high-intensity collisions due to turbulence.

The size distribution data clearly show the dual role of turbulence in the flocculation process. Turbulence promotes the formation of flocs initially, but as the intensity of turbulence increases, it starts to break up the flocs, suggesting that that there is an optimum level of turbulence that results in the maximum size of the flocs. The dual nature of turbulence becomes more evident when the full distributions are examined. In Figure 13, where the full distribution of sediment flocs are presented for bed shear stresses of 0.088, 0.210 and 0.323 Pa, the influence of the turbulence on flocculation is evident.

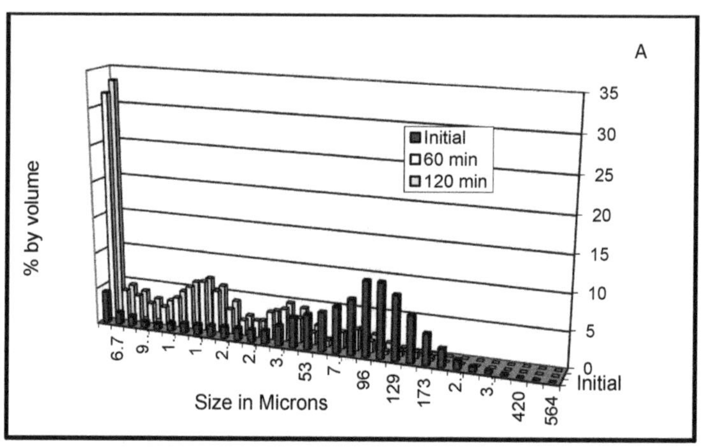
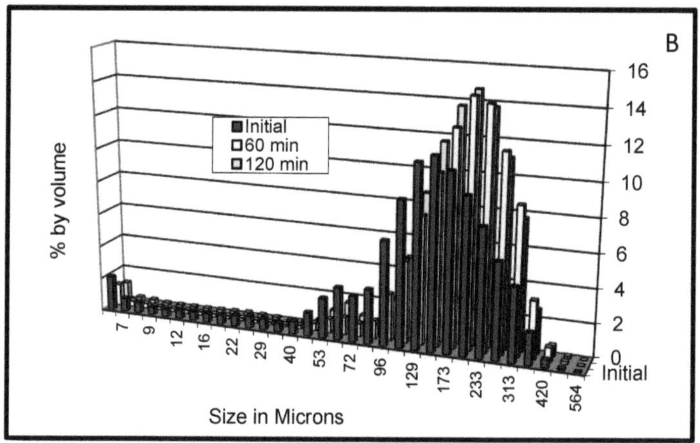
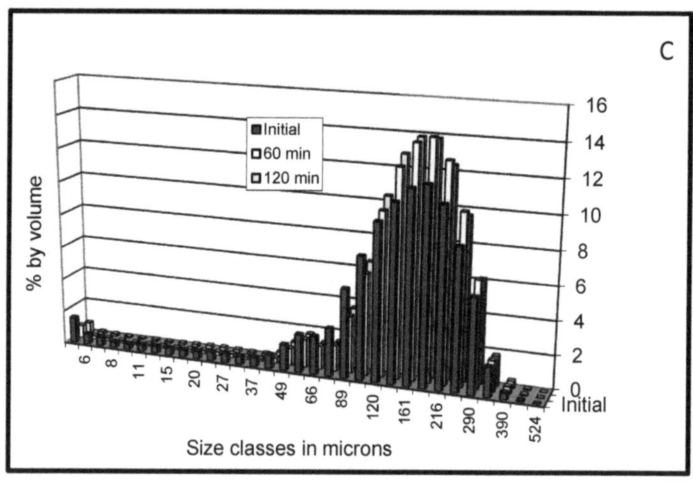

Figure 13. Complete distribution of floc sizes: (**A**) bed shear stress of 0.088 Pa; (**B**) bed shear stress of 0.210 Pa; (**C**) bed shear stress of 0.323 Pa.

In Figure 13A, size distributions are shown for three different elapsed times, namely, zero minutes (initial distribution), 60 min and 120 min. The progressive fining of the sediment particles is evident due to the settlement of coarser fractions. Figure 13B shows the full distributions for the bed shear stress of 0.210 Pa. From Figure 13B, one can clearly see evidence of flocculation. At the elapsed time of 60 min, the percent volume of flocs in the size classes over 150 microns is considerably higher than the initial distribution of flocs in the same size range. This is a clear indication that the flocculation of the sediment particles has occurred. The same thing can also be said for the elapsed time of 120 min. Figure 13C shows the complete distribution of the flocs during deposition at the bed shear stress of 0.323 Pa. From this figure, one can see the evidence of floc breakage. The size of the largest floc that was formed during this experiment is about 390 microns, whereas for the lower bed shear stress experiment (0.210 Pa), the largest flocs formed were of the size of about 420 microns. Moreover, the percent by volume of the largest flocs were also lower for the higher bed shear stress run.

3.5. Determination of Cohesion Parameter β and Empirical Parameters B and C

The RIVFLOC model presented in Section 2.1 was tailored to the rotating circular flume, and the measured data from the rotating circular flume experiments were analyzed using this tailored model. Accordingly, the advection-dispersion equation for the rotating circular flume becomes:

$$\frac{\partial C_k}{\partial t} + w_k \frac{\partial C_k}{\partial z} = \frac{\partial}{\partial z}\left(\Gamma \frac{\partial C_k}{\partial z}\right) \tag{26}$$

where w_k is the settling velocity of the kth size fraction, z is the vertical coordinate, t is the time axis and Γ is the mass transfer coefficient. The flow in the rotating flume was considered to be two dimensional and the longitudinal variation was assumed to be negligible. Therefore, the mass balance equation took the simplified form of Equation (26) and the mass balance was considered only in the vertical direction. Boundary conditions at the sediment–water interface were exactly same as in the RIVFLOC model and the transport of sediment was treated in two stages, i.e., a transport stage and a flocculation stage, as was done in the RIVFLOC model. The flocculation module and the properties of flocs such as density and settling velocity of the sediment flocs were treated exactly the same way as in the RIVFLOC model.

Results from the deposition experiments were analyzed using this model and a comparison between the model predictions and measurements are shown in Figure 14 for three different bed shear stress tests (0.323, 0.164, and 0.123 Pa).

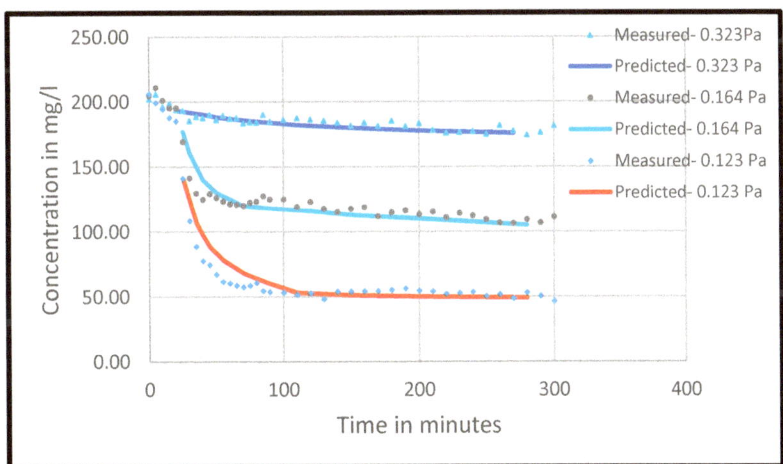

Figure 14. Comparison between model predictions and measured deposition data.

In this analysis, the deposition test with the highest bed shear stress (0.323 Pa) was chosen for calibration, and the coefficients β, b and c were determined as part of the calibration process. The calibration was carried out using a trial-and-error procedure. To start the procedure, the values of b and c as given in Lau and Krishnappan [47] were chosen as the first approximation. Using the values of b and c, and using different values of β, the size distributions of the suspended flocs were determined using the model and the β value that gave the best agreement with the measured distributions was chosen. Using this chosen value of β, a range of b and c values were then tried in the model and the variation of concentration as a function of time was determined. This variation was then compared with the measured data and the set of values that gave the closest match was chosen. Then the process was repeated until the size distribution predictions and the concentration-time variation predictions agreed well with the measured data. Such a convergence was achieved within a limited number of iterations (usually within three). The values obtained for these parameters were as follows: b = 0.02, c = 1.35 and β = 0.010. The predicted concentration–time curves for the other two tests using the calibrated values for the parameters are plotted in Figure 14, and it can be seen from this figure that the agreement is good.

The comparisons of size distributions predicted and measured for the test (0.323 Pa) are shown in Figure 15. Figure 15A is for an elapsed time of 30 min from the start of the deposition experiments. It can be seen from this figure that the agreement between the predicted distributions and measured distribution is reasonable. Figure 15B,C are for elapsed times of 60 and 120 min, and both of these figures show a reasonable agreement between model predictions and the measured size distributions.

Knowing the empirical parameters b and c, density and settling velocity of the sediment flocs can be related to the size of the flocs as shown in Figure 16.

From Figure 16, one can see that the settling velocity of the flocs increases with the size of the flocs when the floc sizes are small, but after reaching a maximum value of about 0.18 mm/s, settling velocity actually decreases with the increase in floc sizes and it approaches zero for larger floc sizes. This is because the effective density is close to that of water for larger flocs. Size distribution data also support this observation. The size distribution of flocs that stayed in suspension were large and the effective density tending towards that of the suspending medium. An inverse relationship between the settling velocity and the size of flocs for larger flocs was also observed by other investigators [61,62].

3.6. Review of the Variability of Model Parameters

The variability of the model parameters among different river systems summarized in Table 2 is reviewed here in Figure 17.

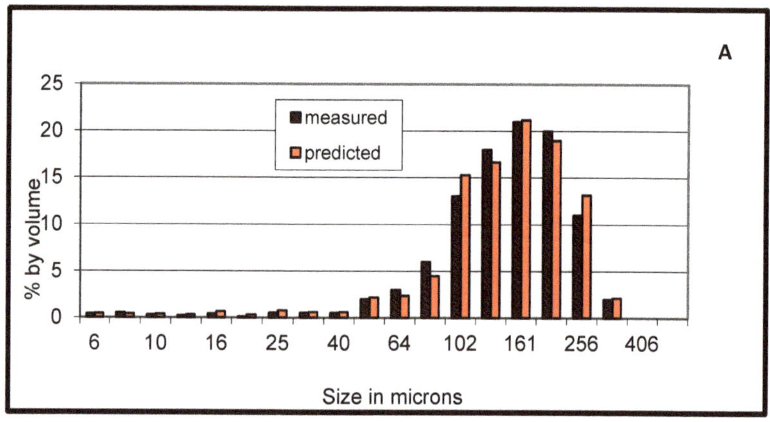

Figure 15. *Cont.*

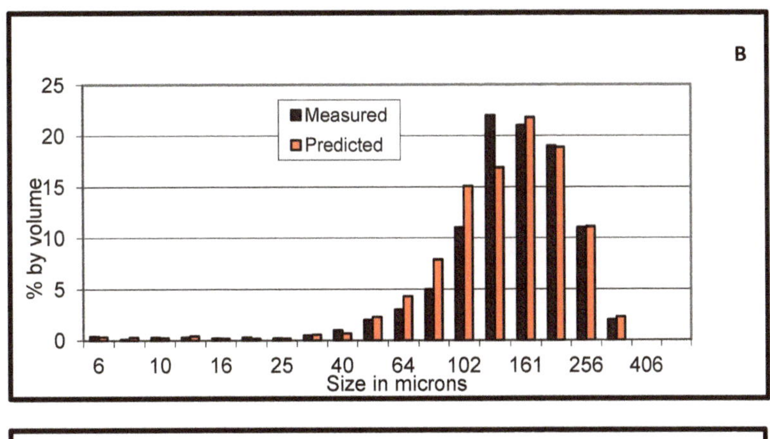

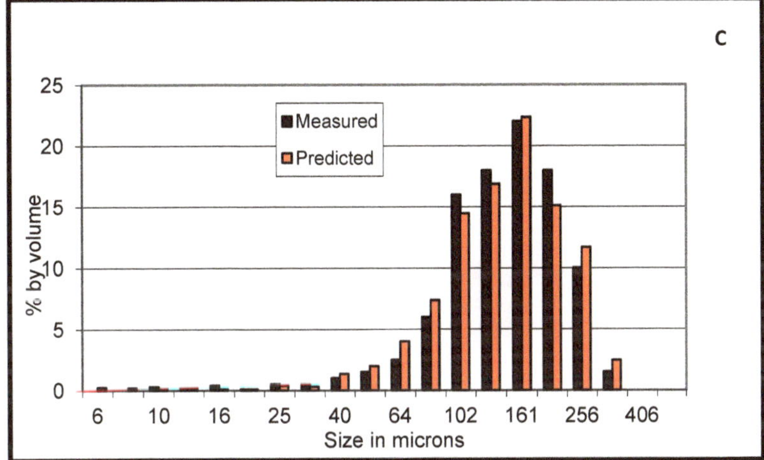

Figure 15. Comparison of measured and predicted size distribution during deposition under the shear stress of 0.323 Pa: (**A**) elapsed time of 30 min; (**B**) elapsed time of 60 min; (**C**) elapsed time of 120 min.

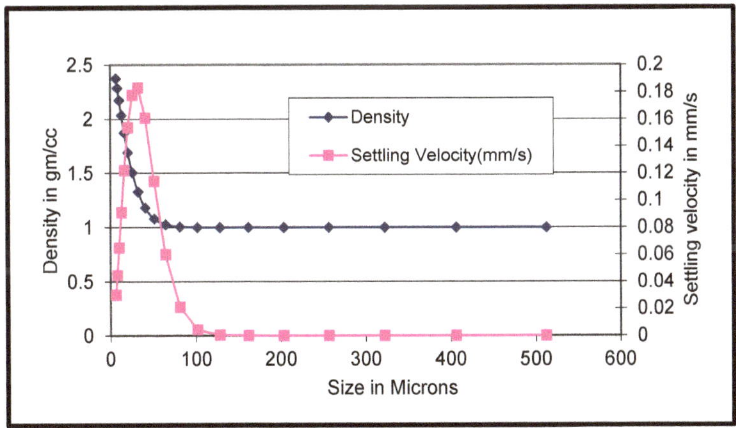

Figure 16. Density and settling velocity of sediment flocs from the Hay River.

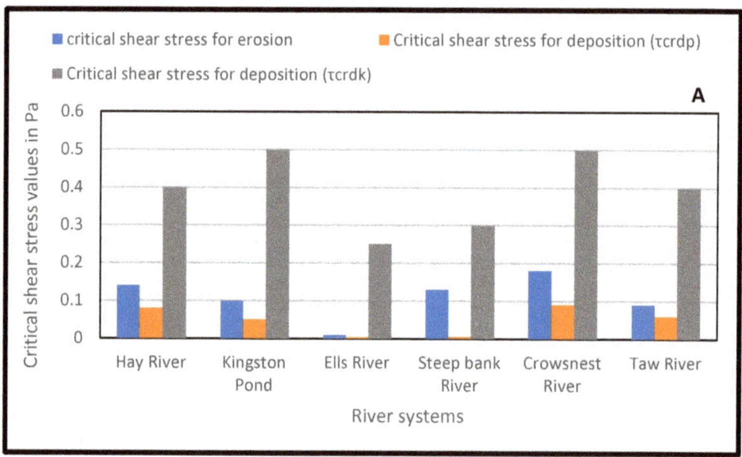

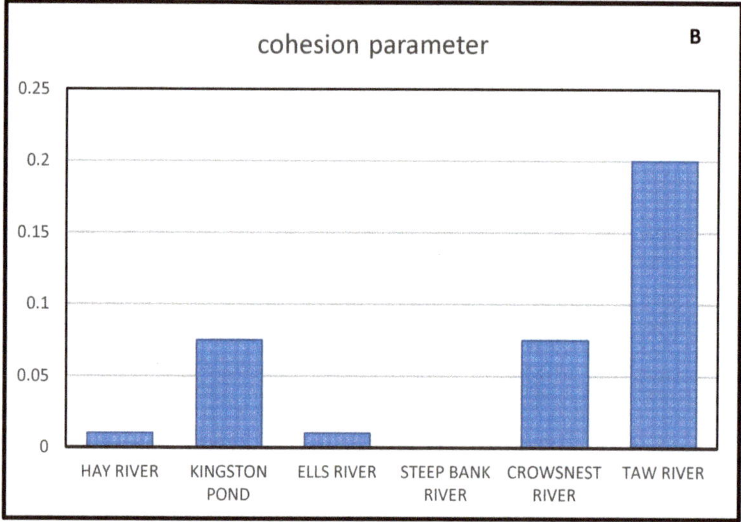

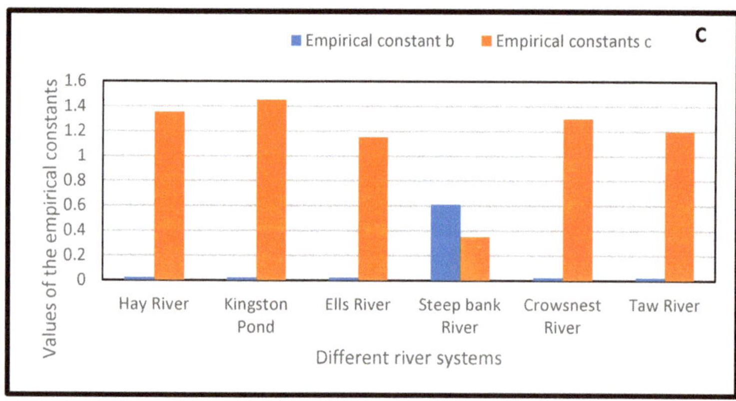

Figure 17. Variability of model parameters among different river systems: (**A**) critical shear stresses for erosion and deposition; (**B**) cohesion parameter; (**C**) empirical constants.

Figure 17A shows the variability of the critical shear stress for erosion and the critical shear stresses for deposition (τ_{crdp}, τ_{crdk}) for all the river systems. From Figure 17A, one can see that the critical shear stress for erosion is always higher than the critical shear stress for deposition (Partheniades definition, τ_{crdp}) because the sediment depositing to the bed gets attached to the sediment in the bed due to cohesion and a higher stress is needed to dislodge the deposited particle. It can also be seen from Table 3 and Figure 17A that there is about 10-fold variation in the values of both critical shear stress for erosion and the critical shear stress for deposition (Partheniades definition, τ_{crdp}) for the river systems. The critical shear stress for deposition (Krones definition, τ_{crdk}) varies from 0.25 to 0.50 Pa and it is much higher than the critical shear stress for deposition (Partheniades definition, τ_{crdp}). The proportions of these three stresses in each river system are different and point to the need for testing site-specific sediments in the flume in order to achieve a reasonable accuracy in model predictions.

Table 3. Maximum settling velocity and optimum floc size for different river systems.

River Systems	Optimum Floc Size in Microns	Maximum Settling Velocity in mm/s
Kingston storm water pond, Canada	20.0	0.08
Taw River, UK	20.0	0.09
Hay River in NWT, Canada	25.0	0.12
Crowsnest River in Canada	32.0	0.15
Ells River in Canada	51.0	0.37
Steepbank River in Canada	n/a	Monotonic increase as a function of floc size

Figure 17B shows the variability of the cohesion parameter, β, for the various river systems. Cohesion parameter is a measure of the efficiency of the system to flocculate for a given number of particle collisions. The variability of the cohesion parameter is in the of range of 0 to 0.1 except for the Taw River system for which the cohesion parameter is 0.20. The high value of the cohesion parameter for the Taw River system is due to the fact that it drains an organic rich peaty and podzolic moorland soils near its source and clayey gley and brown earth soils in more intensely formed lowland adjacent to the moor. Cohesion parameter variability also demonstrates the need for site-specific sediment testing in flumes for reliable predictions from models.

From Figure 17C, one can see that the variability in the empirical constants b and c are not high except for the Steepbank River for which the values of the constants are reversed, suggesting that the relationship between the floc sizes and density is very different for the Steepbank River in comparison to the other river systems. Indeed, if one examines the settling velocity relationship with floc sizes for different river systems, one can see that the relationship for the Steepbank River is very different. The settling velocity vs. the floc size relationships are plotted in Figure 18.

Figure 18 shows that the settling velocity variation as a function of floc sizes is similar for sediment from different river systems except for the Steepbank River. The Steepbank River, which is located in the Athabasca Oil Sands region in Alberta, is impacted by the oil sands development. The river flows through the bituminous soils and hence the sediment in the river is coated with bitumen and consequently becomes hydrophobic. As a result, the sediment flocs are tightly bound and have a higher density and behave like cohesionless sediment without much flocculation. In fact, the cohesion parameter for this sediment is zero. Such variability in modelling parameters also points to the need for testing site-specific sediments to obtain reliable predictions of transport models.

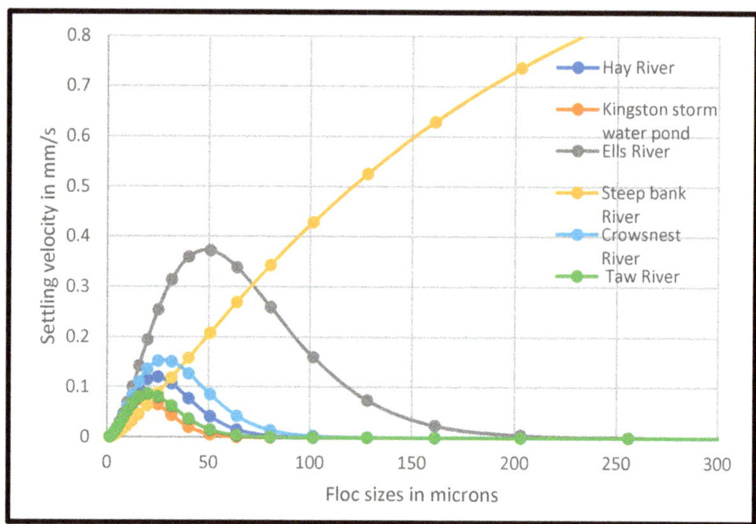

Figure 18. Settling velocity vs. floc size for the different river systems.

3.7. Importance of the Entrapment Process

In most of the river systems for which the RIVFLOC model was applied (i.e., the river systems listed in Table 2), the bed shear stresses were one to two orders of magnitude higher than the critical shear stress for deposition (τ_{crdk}, Krone's definition), and therefore the deposition flux specified to the model is zero (see Equation (3)), which means that the fine sediment entering a river system will be transported as a plug flow without any deposition to the river beds. However, the sampling of the bed materials in some of these rivers does show evidence of fine sediment presence in these beds [63]. Size fractions of fine sediment found in these beds do overlap with the sizes fractions of sediment found in suspension. The presence of fine sediment in the bed is attributed to the process of entrapment in which the fine sediment ingresses into the interstitial voids of the bed sediment as the sediment–water mixture flows through the bed sediment matrix due to infiltration. The cohesion property of the fine sediment causes the fine sediment to stick to the bed matrix [16,17]. Krishnappan and Engel [43] studied the entrapment process using the rotating circular flume and found that the deposition of fine sediment can occur at bed shear stresses much larger than the critical shear stress for deposition, τ_{crdk}.

Figure 19 shows results from the entrapment experiments carried out by Krishnappan and Engel [43]. In these experiments, kaolin was used as a surrogate for fine sediment, and deposition experiments were conducted for two different bed conditions, namely, a plane, impervious bed and a bed covered with 8 mm gravel. Two different bed shear stresses were tested for each configuration (0.165 and 0.325 Pa for impervious runs and 0.30 and 0.70 Pa for gravel bed runs). The concentration variations as a function of time measured for these tests are shown in Figure 19. From Figure 19, one can see that the time variation of concentration profiles of impervious bed runs is significantly different from those of gravel bed runs. Note that when the bed is covered with gravel, almost all of the initially suspended material was entrapped into the voids of the gravel and the concentration of sediment in suspension was nearly zero. When the bed was impervious (smooth flume bed), the sediment stayed mainly in suspension and the deposition rates were considerably smaller (judging by the slope of the concentration decay curves). The bed shear stress value of 0.70 Pa is higher than the critical shear stress for the deposition of kaolin (τ_{crdk}), which was estimated to 0.50 Pa.

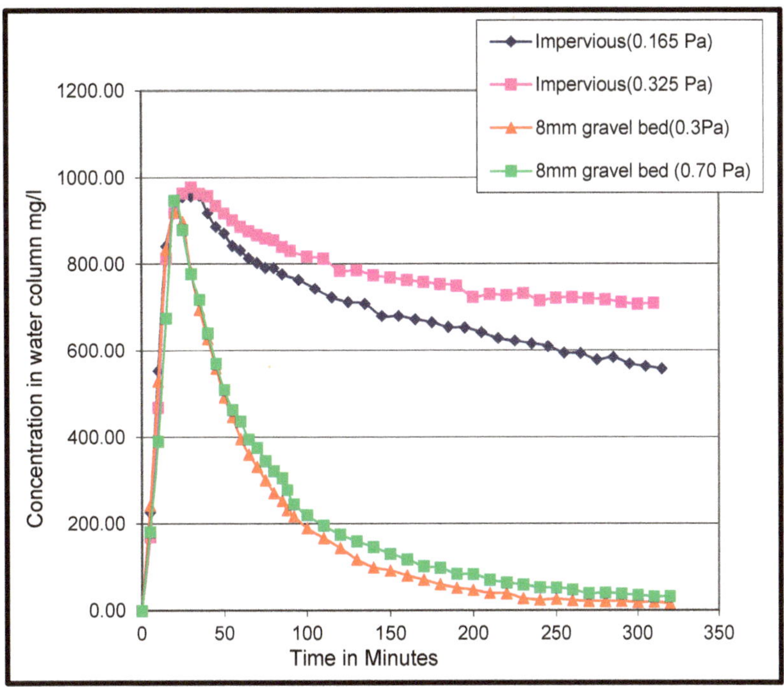

Figure 19. Summary of entrapment experiments in the rotating circular flume.

Krishnappan and Engel [43] modelled the entrapment of kaolin in the gravel bed using the RIVFLOC model tailored to the rotating flume and found that the coefficient of entrapment varied from 0.19 to 0.94. No attempt was made in that study to relate the coefficient of entrapment to bulk properties of the bed such as the porosity of the gravel substrate, thickness of the gravel bed and the permeability.

Glasbergen [64] studied the entrapment process using the rotating circular flume for fine sediment and coarse gravel from the Elbow River in Alberta. He evaluated the coefficient of entrapment using the method of Krishnappan and Engel [43] and obtained a value of 0.20 for the coefficient. He too did not make any attempt to relate the coefficient of entrapment to the bulk properties of the river-bed materials. Therefore, there is a need for further research in this area to have a better understanding of the entrapment process, which is crucial for modelling the fate of the fine sediments and the associated contaminants.

Fine sediment that is entrapped in coarse gravel beds remain in the bed until the coarse sediment beds are eroded, which is likely under high flows. The initiation of the movement of the coarse gravel can be predicted reliably using the Shield's diagram for critical shear stress for the erosion of coarse-grained sediment.

4. Discussion

The model parameters obtained by the application of the RIVFLOC model to the rotating flume experiments for different river systems lead to some interesting observations. For example, the cohesion parameter obtained for the Taw River system in the UK is about two to three times higher than the values obtained for the other river systems. As indicated earlier, cohesion parameter reflects the efficiency of collisions in the formation of flocs. This means that, for the Taw River sediment, more collisions will result in the formation of flocs. This can be attributed to the organic matter content of the sediment. Indeed, Taw River flows through an organic rich peaty and podzolic moorland near its source and an intensely farmed lowland downstream.

The empirical constants determined for different river systems enable the calculation of settling velocities as a function of floc sizes as shown in Figure 18. From this figure, the maximum settling velocity and the flocs size corresponding to the maximum settling velocity (optimum floc size) are extracted and summarized in Table 3.

From Table 3, one can see that the Ells River sediment has the maximum settling velocity of 0.37 mm/sec and the optimum floc size of 51 microns compared to all other sediments. It is interesting to note that the Taw River sediment, which has the highest cohesion parameter, is not the one with the highest settling velocity and the largest optimum floc size. Efficiency in flocculation does not imply that the resulting flocs will have the largest floc size and the highest settling velocity. The high settling velocity and the large optimum floc size of the Ells River may be attributed to the river basin, which is a tar-sand region with a high percentage of bituminous coal content.

The lowest settling velocity and the smallest optimum floc size corresponds to the sediment in the Kingston storm water pond, which is an in-stream pond with lower bed shear stresses over a large area of the pond. There is substantial deposition of sediment in the pond and the sediment in suspension are finer with lower settling velocities.

The Steepbank River sediments show an interesting behavior. The sediment in this river behaves as a cohesionless sediment with practically no flocculation (the cohesion parameter for this sediment is zero) and the settling velocity shows a monotonic increase with the size of the sediment. It should be noted that the Steepbank River basin is also in the tar sand region, but it is highly impacted by the operation of the Fort McMurray tar sands oil development project. The Ells River basin, on the other hand, is not impacted by this project.

One of the major difficulties in carrying out the laboratory investigation of the cohesive sediment transport processes involves the transportation of a large volume of river water and sediment mixture (about 500 to 1000 L) from the river site to the laboratory flume. For river sites in Canada, it was possible to transport such large volume of samples in transport trucks fitted with refrigerated storage containers. For the Taw River study, river water was not used because of the cost of transportation, and hence the local tap water was used to carry out the tests. The impact of the tap water use instead of river water was not addressed in this study.

Another source of error that might have had an effect on the sediment behavior is the lack of temperature control for the rotating flume. The rotating flume experiments were carried out in room temperature after the sediment–water mixtures were placed in the flume. Prior to that, the samples were stored in the refrigerator maintained at 4 degrees centigrade. Lack of temperature control could have had an effect on the bacterial growth during the conduct of experiments in the rotating flume.

Error estimates in the determination of model parameters were not made in the reviewed studies. There is a degree of human judgement involved in the determination of the critical conditions at which the erosion and deposition of sediment is occurring and how the calculated distributions agree with the measured distributions to determine the empirical constants and cohesion parameter. The methods employed in these studies were aimed to reduce the human judgement in establishing these model parameters. The measurement of sediment concentrations and bed shear stresses were also carried out with a high degree of precision (within ± 1 mg/L and ± 0.001 Pa).

5. Conclusions

A semi-empirical modelling technique for the transport of cohesive sediment is needed because of the large number of controlling parameters governing the erosion and deposition of cohesive sediments and the flocculation process that distinguishes the fine sediment from the cohesionless coarse-grained sediment for which there are well established models available in the literature. An examination of the details of a semi-empirical cohesive sediment transport model called RIVFLOC highlights the model parameters that needed to be determined using a rotating circular flume. The parameters that were determined using the rotating circular flume include the critical shear stress for erosion of the cohesive sediment, critical shear stress for deposition according to the definition of Partheniades, critical shear stress for deposition according to the definition of Krone, the cohesion parameter governing the flocculation of cohesive sediment and two empirical parameters that define the density of the floc in terms of the size of the flocs.

The critical shear stress for the erosion of cohesive sediment is different from the critical shear stress for deposition, suggesting that the cohesive sediments do not undergo erosion and deposition simultaneously, unlike the cohesionless sediment for which the two critical conditions coincide and the sediment undergoes simultaneous erosion and deposition. The flocculation of the cohesive sediment was inferred from the measurement of the size distribution of sediment flocs using a sizing instrument, which shows that the cohesive sediment did indeed flocculate. The density of the flocs is dependent on the size of the flocs; as the size of the floc increases, more of the surrounding fluid is incorporated in the structure of the floc. An empirical equation relates the floc density with floc size, which in turn provides a relationship for the settling velocity based on the Stokes equation.

The semi-empirical RIVFLOC model was applied to a number of river systems to model the transport of cohesive sediments and associated contaminants. The various model parameters determined for these river systems were compared and the variability in the model parameters were examined. The variabilities in these parameters suggest that the establishment of model parameters for site-specific sediment is important for accurate predictions of the cohesive sediment transport models.

Application of the RIVFLOC model to various river systems points to the need for modelling the entrapment process of cohesive sediment transport. The entrapment process is responsible for the presence of fine sediment in coarse-grained sediment bed matrix in gravel bed rivers where the bed shear stress far exceeds the critical shear stress for deposition, as defined by Krone. More research is needed to relate the coefficient of entrapment with the bulk properties of coarse-grained sediment bed matrix.

Funding: This research received no external funding.

Institutional Review Board Statement: This study was conducted in accordance with the Declaration of Helsinki.

Informed Consent Statement: Not applicable.

Data Availability Statement: Data supporting reported results can be requested from the author.

Conflicts of Interest: The authors declare no conflict of interest.

References

1. Raudkivi, A.J. *Loose Boundary Hydraulics*, 4th ed.; Balkema, A.A., Ed.; CRC Press: Boca Raton, FL, USA.
2. Van Leussen, W. Aggregation of Particles, Settling Velocity of Mud Flocs. In *Physical Processes in Estuaries*; Van Leussen, D., Ed.; Springer: New York, NY, USA, 1988.
3. Partheniades, E. A study of erosion and deposition of cohesive soils in salt water. Ph.D. Thesis, University of Florida, Gainesville, FL, USA, 1962.
4. Partheniades, E. Results of recent investigation on erosion and deposition of cohesive sediments. In *Sedimentation*; Shen, H.W., Ed.; Colorado State University: Fort Collins, CO, USA, 1972.
5. Partheniades, E. The present state of knowledge and needs for future research on cohesive sediment dynamics. In *River Sedimentation, Proceedings of the 3rd International Symposium*; Wang, S.Y., Shen, H.W., Ding, L.Z., Eds.; School of Engineering, University of Mississippi: Jackson, MS, USA, 1986; Volume 3, pp. 3–25.
6. Partheniades, E.; Kennedy, J.F. Depositional behaviour of fine sediments in a turbulent fluid motion. In Proceedings of the 10th Conference on Coastal Engineering, London, UK, 1–5 September 1966; American Society of Civil Engineers: New York, NY, USA, 1966; Volume 2, pp. 707–724.
7. Patheniades, E.; Cross, R.H.; Arora, A. Further results on the deposition of cohesive sediments. In Proceedings of the 11th Conference on Coastal Engineering, London, UK, 16–20 September 1968; American Society of Civil Engineers: New York, NY, USA, 1968; Volume 2, pp. 723–742.

8. Einstein, H.A.; Anderson, A.G.; Johnson, J.W. A distinction between bed-load and suspended load in natural streams. *Trans. Am. Geophys. Union* **1940**, *21*, 628–633. [CrossRef]
9. Einstein, H.A. *The Bed Load Function for Sediment Transportation in Open Channel Flows*; Technical Bulletin No. 1026; U.S. Department of Agriculture, Soils Conservation Service: Washington, DC, USA, 1950.
10. Yalin, M.S. *Mechanics of Sediment Transport*, 1st ed.; Pergamon Press: Braunschweig, Germany, 1972.
11. Krone, R.B. *Flume Studies of Transport of Sediment in Estuarial Shoaling Processes: Final Report*; Hydraulic Engineering Laboratory and Sanitary Engineering Research Laboratory, University of California: Berkeley, CA, USA, 1962.
12. Mehta, A.J.; Partheniades, E. An Investigation of the Depositional Properties of Flocculated Fine Sediments. *J. Hydraul. Res.* **1975**, *13*, 361–381. [CrossRef]
13. Lick, W. Entrainment, deposition and transport of fine grained sediments in Lakes. *Hydrobiologia* **1982**, *91*, 31–40. [CrossRef]
14. Krishnappan, B.G.; Engel, P. Critical shear stresses for erosion and deposition of suspended sediments of the Fraser River. In *Cohesive Sediments*; Burt, N., Parker, R., Watts, J., Eds.; John Wiley and Sons: Chichester, UK, 1997; pp. 279–288.
15. Krishnappan, B.G.; Marsalek, J. Chapter 8: An example of modelling flocculation in a freshwater aquatic system. In *Flocculation in Natural and Engineered Environmental Systems*; Droppo, I., Leppard, G.G., Liss, S.N., Mulligan, T.G., Eds.; CRC Press: Boca Raton, FL, USA, 2005; pp. 171–188.
16. Packman, A.I.; Brooks, N.H.; Morgan, J.J. Physico-chemical model for colloid exchange between a stream and a sand stream bed with bedforms. *Water Resour. Res.* **2000**, *36*, 2351–2361. [CrossRef]
17. Packman, A.I.; Brooks, N.H.; Morgan, J.J. Kaolinite exchange between a stream and stream bed: Laboratory experiment and validation of a colloidal transport model. *Water Resour. Res.* **2000**, *36*, 2363–2372. [CrossRef]
18. Owen, M.W. *A Detailed Study of Settling Velocities of an Estuarine Mud*; Report INT78; Hydraulic Research Station: Wallingford, UK, 1970.
19. Ongley, E.D.; Bynoe, M.C.; Percival, J.B. Physical and geochemical characteristics of suspended solids, Wilton Creek, Ontario. *Can. J. Earth Sci.* **1981**, *18*, 1365–1379. [CrossRef]
20. Droppo, I.G.; Ongley, E.D. Flocculation of suspended solids in Southern Ontario rivers. *Water Res.* **1994**, *28*, 1799–1809. [CrossRef]
21. Droppo, I.G.; Leppard, G.G.; Liss, S.N.; Milligan, T.G. Chapter 20: Opportunities, Needs and Strategic Direction for Research on Flocculation in Natural and Engineered Systems. In *Flocculation in Natural and Engineered Environmental Systems*; Droppo, I., Leppard, G.G., Liss, S.N., Mulligan, T.G., Eds.; CRC Press: Boca Raton, FL, USA, 2005; pp. 407–421.
22. Rao, S.S.; Droppo, I.G.; Taylor, C.M.; Burnison, B.K. *Fresh Water Bacterial Aggregate Development: Effects of Dissolved Organic Matter*; NWRI Contribution 91–75; National Water Research Institute, CCIW: Burlington, ON, Canada, 1991.
23. Leppard, G.G. Evaluation of electron microscope techniques for the description of aquatic colloids. In *Environmental Particles*; Buffle, J., Van Leeuwen, H.P., Eds.; IUPAC Environmental and Analytical Chemistry Series; Lewis Publishers: Boca Raton, FL, USA, 1992; Volume 1, pp. 231–289.
24. Ruehrwein, R.A.; Ward, D.W. Mechanism of clay aggregation by polyelectrolytes. *Soil Sci.* **1952**, *73*, 485–492. [CrossRef]
25. La Mer, V.K.; Heley, T.W. Adsorption-flocculation reactions of macromolecules at the solid-liquid interfaces. *Rev. Pure Appl. Chem.* **1963**, *13*, 112–132.
26. Busch, P.L.; Stumm, W. Chemical interactions in the aggregation of bacteria: Bio-flocculation in waste treatment. *Environ. Sci. Technol.* **1968**, *2*, 49–53. [CrossRef]
27. Ariathurai, R.; Arulanandan, K. Erosion rates of cohesive soils. *J. Hydraul. Div. ASCE* **1978**, *104*, 279–283. [CrossRef]
28. Lee, D.-Y.; Lick, W.; Kang, S.W. The Entrainment and Deposition of Fine-Grained Sediments in Lake Erie. *J. Great Lakes Res.* **1981**, *7*, 224–233. [CrossRef]
29. Dixit, J.G. Resuspension Potential of Deposited Kaolinite Beds. Master's Thesis, University of Florida, Gainesville, FL, USA, 1982.
30. Parchure, T.M. Erosional Behaviour of Deposited Cohesive Sediments. Ph.D. Thesis, University of Florida, Gainesville, FL, USA, 1984.
31. Krishnappan, B.G. Erosion behaviour of fine sediment deposits. *Can. J. Civ. Eng.* **2004**, *31*, 759–766. [CrossRef]
32. Hayter, E.J. *Finite Element Hydrodynamic and Cohesive Sediment Transport Modelling System*; Department of Civil Engineering, Clemson University: Clemson, SC, USA, 1987.
33. Krishnappan, B.G. Modelling of cohesive sediment transport. In Proceedings of the International Symposium on the Transport of Suspended Sediment and its Mathematical Modelling, Florence, Italy, 2–5 September 1991.
34. Krishnappan, B.G. Cohesive sediment transport studies using a rotating circular flume. In Proceedings of the 7th International Conference on Hydro-Science and Engineering (ICHE-2006), Philadelphia, PA, USA, 10–13 September 2006.
35. Krishnappan, B.G.; Marsalek, J. Modelling of flocculation and transport of cohesive sediment from an on-stream stormwater detention pond. *Water Res.* **2002**, *36*, 3849–3859. [CrossRef]
36. Droppo, I.G.; Krishnappan, B.G. Modeling of hydrophobic cohesive sediment transport in the Ells River Alberta, Canada. *J. Soils Sediments* **2016**, *16*, 2753–2765. [CrossRef]
37. Krishnappan, B.G. *Personal Communication*; Krishnappan Environmental Consultant: Hamilton, ON, Canada, 2015.
38. Stone, M.; Krishnappan, B.G.; Silins, U.; Emelko, M.B.; Williams, C.H.S.; Collins, A.L.; Spencer, S.A. A new framework for modelling fine sediment transport in rivers includes flocculation to inform reservoir management in wildfire impacted watersheds. *Water* **2021**, *13*, 2319. [CrossRef]

39. Stone, M.; Krishnappan, B.G.; Granger, S.; Upadhayay, H.R.; Zhang, Y.; Chivers, C.A.; Decent, Q.; Collins, A.L. Deposition behaviour of cohesive sediment in the upper Taw observatory, Southern UK: Implications for management and modelling. *J. Hydrol.* **2021**, *598*, 126145. [CrossRef]
40. Yotsukura, N.; Sayre, W.W. Transverse mixing in natural channels. *Water Resour. Res.* **1976**, *12*, 695–704. [CrossRef]
41. Mehta, A.J.; Partheniades, E. Resuspension of deposited cohesive sediment beds. In Proceedings of the 18th Coastal Engineering Conference, Cape town, South Africa, 14–19 November 1982; Edge, B.L., Ed.; ASCE: New York, NY, USA, 1983; Volume 2, pp. 1569–1588.
42. Krishnappan, B.G.; Stone, M.; Granger, S.J.; Upadhayay, H.R.; Tang, Q.; Zhang, Y.; Collins, A.L. Experimental investigation of erosion characteristics of fine grained cohesive sediments. *Water* **2020**, *12*, 1511. [CrossRef]
43. Krishnappan, B.G.; Engel, P. Entrapment of fines in coarse sediment beds. In *River Flow 2006*; Ferrira, A., Leal, C., Eds.; Taylor and Francis Group: London, UK, 2006; pp. 817–824.
44. Stone, H.L.; Brian, P.L.T. Numerical solution of convective transport problems. *Am. Inst. Chem Eng. J.* **1963**, *9*, 681–688. [CrossRef]
45. Krishnappan, B.G.; Lau, Y.L. *RIVMIX MK2: User Manual*; Hydraulics Division, National Water Research Institute: Burlington, ON, Canada, 1985; p. 25.
46. Fuchs, N.A. *Mechanics of Aerosols*; Pergamon: New York, NY, USA, 1964; p. 408.
47. Lau, Y.L.; Krishnappan, B.G. *Measurement of Size Distribution of Settling Flocs*; NWRI Contribution No. 97–223; Environment Canada: Burlington, ON, Canada, 1997; p. 21.
48. Valioulis, I.A.; List, E.J. Numerical simulation of a sedimentation Basin. 1. Model Development. *Environ. Sci. Technol.* **1984**, *18*, 242–247. [CrossRef]
49. Tambo, N.; Watanabe, Y. Physical aspects of flocculation process—I: Fundamental Treatise. *Water Res.* **1979**, *13*, 429–439. [CrossRef]
50. Djordjevic, S. Mathematical model of unsteady transport and its experimental verification in a compound open channel flow. *J. Hydraul. Res.* **1993**, *31*, 229–248. [CrossRef]
51. Guan, Y.; Altinakar, M.S.; Krishnappan, B.G. Modelling of lateral flow distribution in compound channels. In Proceedings of the River Flow 2002 International Conference on Fluvial Hydraulics, Louvan-la Neuve, Belgium, 4–6 September 2002.
52. Krishnappan, B.G. *User Manual: Unsteady, Non-Uniform Mobile Boundary Flow Model—MOBED*; Hydraulics Division, National Water Research Institute: Burlington, ON, Canada, 1981; p. 197.
53. Krishnappan, B.G. *MOBED User Manual Update I.*; Hydraulics Division, National Water Research Institute: Burlingtom, ON, Canada, 1983; p. 72.
54. Krishnappan, B.G. *MOBED User Manual Update II.*; Hydraulics Division, National Water Research Institute: Burlingtom, ON, Canada, 1986; p. 95.
55. Krishnappan, B.G. Rotating Circular Flume. *J. Hydraul. Eng. ASCE* **1993**, *119*, 658–667. [CrossRef]
56. Krishnappan, B.G.; Engel, P.; Stephens, R. *Shear Velocity Distribution in a Rotating Flume*; NWRI Contribution No. 94–102; NWRI: Burlington, ON, Canada, 1994.
57. Petersen, O.; Krishnappan, B.G. Measurement and analysis of flow characteristics in a rotating circular flume. *J. Hydraul. Res. IAHR* **1994**, *32*, 483–494. [CrossRef]
58. Krishnappan, B.G.; Engel, P. Distribution of bed shea stress in a rotating circular flume. *J. Hydraul. Eng. ASCE* **2004**, *130*, 324–331. [CrossRef]
59. Rosten, H.I.; Spalding, D.B. *The PHOENICS Reference Manual, TR/200*; CHAM Ltd.: London, UK, 1984.
60. Weiner, B.B. Particle and droplet sizing using Fraunhofer Diffraction. In *Modern Method of Particle Size Analysis*; Bath, H.G., Ed.; John Wiley and Sons: London, UK, 1984; Chapter 5.
61. Petticrew, E.L. The composite nature of suspended and gravel stored fine sediments in streams: A case study of O'Ne-eil Creek, British Columbia, Canada. In *Flocculation in Natural and Engineered Environmental Systems*; Droppo, I.G., Leppard, G.G., Liss, S.N., Milligan, T.G., Eds.; CRC Press: Boca Raton, FL, USA, 2005; pp. 71–93.
62. Atkinson, J.F.; Chakraborti, R.K.; VanBenschoten, J.E. Effect of floc size and shape in particle aggregation. In *Flocculation in Natural and Engineered Environmental Systems*; Droppo, I.G., Leppard, G.G., Liss, S.N., Milligan, T.G., Eds.; CRC Press: Boca Raton, FL, USA, 2005; pp. 95–120.
63. Suzanne, C.L. Effects of Natural and Anthropogenic Non-Point Source Disturbances on the Structure and Function of Tributary Ecosystems in the Athabasca Oil Sands Region. Master's Thesis, University of Victoria, Victoria, BC, Canada, 2015.
64. Glasbergen, K.A. The Effect of Coarse Gravel on Cohesive Sediment Entrapment. Master's Thesis, University of Waterloo, Waterloo, ON, Canada, 2013.

Review

Effects of River-Ice Breakup on Sediment Transport and Implications to Stream Environments: A Review

Spyros Beltaos [1,*] and Brian C. Burrell [2]

1   Watershed Hydrology and Ecology Research Division, Environment and Climate Change Canada, Canada Centre for Inland Waters, 867 Lakeshore Rd, Burlington, ON L7S 1A1, Canada
2   Independent Researcher, P.O. Box 3027, Fredericton, NB E3A 5G8, Canada; bburrell@nbnet.nb.ca
*   Correspondence: spyros.beltaos@canada.ca

Abstract: During the breakup of river ice covers, a greater potential for erosion occurs due to rising discharge and moving ice and the highly dynamic waves that form upon ice-jam release. Consequently, suspended-sediment concentrations can increase sharply and peak before the arrival of the peak flow. Large spikes in sediment concentrations occasionally occur during the passage of sharp waves resulting from releases of upstream ice jams and the ensuing ice runs. This is important, as river form and function (both geomorphologic and ecological) depend upon sediment erosion and deposition. Yet, sediment monitoring programs often overlook the higher suspended-sediment concentrations and loads that occur during the breakup period owing to data-collection difficulties in the presence of moving ice and ice jams. In this review paper, we introduce basics of river sediment erosion and transport and of relevant phenomena that occur during the breakup of river ice. Datasets of varying volume and detail on measured and inferred suspended-sediment concentrations during the breakup period on different rivers are reviewed and compared. Possible effects of river characteristics on seasonal sediment supply are discussed, and the implications of increased sediment supply are reviewed based on seasonal comparisons The paper also reviews the environmental significance of increased sediment supply both on water quality and ecosystem functionality.

Keywords: aquatic life; bank erosion; bed shear; ice jam; sediment pulse; suspended-sediment concentration; water quality

Citation: Beltaos, S.; Burrell, B.C. Effects of River Ice Breakup on Sediment Transport and Implications to Stream Environments: A Review. *Water* **2021**, *13*, 2541. https://doi.org/10.3390/w13182541

Academic Editors: Bommanna Krishnappan and Achim A. Beylich

Received: 28 July 2021
Accepted: 9 September 2021
Published: 16 September 2021

Publisher's Note: MDPI stays neutral with regard to jurisdictional claims in published maps and institutional affiliations.

1. Introduction

Ice formation, growth, and breakup affect the flow dynamics of a river or stream and thus its potential for erosion and transport of sediment [1]. During freeze up, anchor ice release can remove material from the bed of a river or stream. As an ice sheet forms, the resistance to flow created by the ice cover modifies the velocity profile such as to alter flow conveyance and sediment transport. In general, erosion and the transport of sediment decrease as flow and flow velocity decrease during the winter months. However, if the flow is channelized by slush deposits under a solid ice sheet, or an ice cover is uneven (such as a reconsolidated ice cover after midwinter jamming), significant erosion and transport of sediment can occur. Nonetheless, the breakup period, characterized by increased water levels and flows, is generally more important with respect to erosion and sediment transport.

Ice breakup can be described as thermal or mechanical [2]. "Mechanical" breakup is generally a dynamic event associated with rising water levels and ice movement. Rising water levels result from increased liquid precipitation and (or) snowmelt runoff that increases the flow discharge in a river and stream and that also brings particles of fine sediment from the land surface to the watercourse. As water levels rise, the ice cover detaches from the banks, and large segments of the cover move downstream due primarily to tractive forces of flowing water on the under surface of the cover and the downslope component of the weight of ice fragments. As fragments of the ice sheet push against the river banks, they break down into smaller pieces. The scraping and or gouging of river banks by moving ice and the release of sediment during the thawing of the river bank increase sediment concentrations in the river or stream. The potential for increases in sediment concentrations is thus greater during the dynamic transient period of ice breakup than the rather less eventful midwinter and summer low-flow periods. "Thermal" breakup occurs when winter ice undergoes thermal melting and decay, thereby losing a large part of its mechanical strength and thickness before being dislodged. Relative to the more dynamic mechanical events, thermal events have limited potential for erosion and sediment transport; therefore, reference to breakup will, in the following text, refer to mechanical events unless otherwise specified.

Breakup processes can generate significant contributions to the annual sediment load, sometimes with detrimental consequences. The effects of breakup ice jams on flow conveyance and direction can increase or lessen flow velocity and tractive force, thereby leading to sediment removal or deposition, with potential localized detrimental effects with respect to the functioning of water intakes and the conservancy of aquatic habitat. Bed scour under an ice jam can have important implications with respect to pipeline and bridge pier design, as these structures can be damaged or destroyed if undermined. Javes (ice-jam-release waves) exemplify the extreme power and erosive capacity of breakup in their heights, rates of water level rise, celerities, and high speeds of associated ice runs [3]. A sharp rise in suspended-sediment concentrations occurs before the arrival of the ice run, as the leading edge of a jave propagates faster than the main body of released water. Short-duration, jave-related sediment pulses are extremely high concentrations of suspended sediment that contribute sizeable sediment loads and create a potential threat to aquatic life.

Winter ice regimes are changing in response to climatic change [4,5]. As ice processes on rivers change, e.g., greater occurrence of thermal or midwinter breakups, ice action against channel boundaries may be altered, thereby affecting channel geometry and the supply, transport, and deposition of sediment and resulting in geomorphic and ecosystem changes. The annual riverine sediment load may increase due to increased discharges and be redistributed within the seasons, with a larger percentage of the annual load during the winter [6].

The purpose of this paper is to provide a review of the effects of river-ice breakup on sediment concentrations, especially suspended sediment and its implications for water quality and aquatic life. Relevant scientific and technical background literature on sediment erosion and transport is reviewed briefly in the following section. The paper then presents an overview of ice-breakup processes followed by a summary of datasets obtained during the breakup period by several researchers, including detailed measurements in the International Saint John River (NB, Canada, and Maine, USA) and the Athabasca River (Alberta, Canada). Implications for water quality and aquatic life are discussed. In the following text, the terms "concentration" or "sediment concentration" will refer to the concentration of suspended sediment, unless specified otherwise.

2. Sediment Erosion and Transport

2.1. Bank and Bed Erosion Potential

Chassiot et al. [7] identifies several terrestrial and fluvial processes that affect bank erosion. Terrestrial processes include slow-acting preparatory processes that affect ground stability, such as groundwater seepage, freeze-thaw processes, and desiccation cycles, and more rapid, mass-wasting processes, such as debris falls and slumps. Fluvial processes include abrasive removal of bank sediment by flowing water, waves, moving ice, and highly dynamic processes associated with ice-jam release [7]. In permafrost regions and cold regions with seasonally frozen ground, progressive thawing of riverbanks in contact with flowing water results in heat transfer between water and ice that reduces particle attachment (thermal erosion), producing easily removable sediments that can be destabilized and removed by terrestrial and fluvial processes [7].

The occurrence and magnitude of bank erosion depends upon several hydraulic factors, such as flow energy, wave action, and the presence of bank fast ice, and several non-hydraulic factors, such as bank morphology, frost action, vegetation cover, and the effects of human and animal action [7]. The amount of bank erosion thus depends upon the magnitude of the forces acting on the sediment particles in a river bank and the mechanical properties of the soils, such as sediment structure, size distribution, bulk density, and mineral content, and upon the protection provided by vegetation [8].

Vandermause [9] investigated the role played by dynamic ice-breakup events in bank erosion of the Susitna River (Alaska, USA). Extensive field measurements and observations indicated that most of the bank erosion, 54 to 61 percent by sub-reach, occurs or is initiated during dynamic breakup of the winter ice cover. High water discharge in combination with ice floes and ice rubble are dominant erosion factors. Vegetated bars and terrace margins were the most susceptible to erosion, caused by impacting ice floes. Less susceptible were banks adjoining floodplain surfaces and partly protected by vegetation root mats and by shear walls of ice rubble. Bank erosion occurred less in predominantly single-channel reaches than in predominantly multi-channel reaches.

The effects of moving river ice on channel erosion and alignment have been reported, with many channels exhibiting effects of ice on channel morphology [10]. Along the gravel-bed Susitna River, Alaska, consecutive datasets of aerial photography were used to detect the location, magnitude, and timing of large-scale channel change owing to water flow and river-ice processes [10,11]. Vandermause et al. [10] examins a 2011–2015 dataset from the Susitna's Middle River segment to quantify the extent to which bank erosion along the river is driven by river-ice or open-water fluvial processes. To determine which geomorphic surfaces and bank profiles may be more susceptible to erosion, eroded areas are examined to relate erosion extent to bank resistance, which is based on sediment composition, bank height, and vegetation. In reaches 6 and 7 of the middle stretch of the Susitna River, 41% and 7%, respectively, of the erosion occurred during a 5-day breakup period; 20% and 46%, respectively, during the start of the ice breakup though the fall period; and about 36% and 46% of the erosion could not be assigned to a timeframe during the year based on visual evidence [10]. Fluvial processes in general did not appear to cause large-scale bank retreat, although open-water erosional processes contribute to the undercutting of root-reinforced cantilevered banks by entraining non-cohesive sand and silt sediments once water levels are above the basal gravel layer [10].

Three ice-driven erosional processes were identified from field observations and analysis of aerial photography and video along the Susitna River:

1. Abrasion, gouging, and removal by ice rubble of the upper layer of cantilevered rootmats thereby exposing the bank to subsequent fluvial entrainment of fine-grained material and disruption of the armour pavement of cobbles along the bank toe, with rubble-ice impact observed to be more severe for banks near the upstream tip of an island or bar or where flow directed rubble directly against a bank;
2. Gouging due to the impact of moving ice floes on river banks and overtopping and bulldozing of vegetation on low islands and bars, sometimes creating pathways for flow to drain into secondary channels within multi-channel reaches;
3. Increased flow velocity and the removal of armour pavement of cobbles and gravel due to ice congestion in the channel, which could occur because of an ice cover and anchor ice, flow concentration beneath and around an ice jam, and from an ice-jam-release jave. The primary effects of increased shear stress on bank erosion are mobilization of bank toe pavement (most likely due to javes), entrainment of bank material, and undercutting and failure of the vegetated top of bank [10].

Four factors that mitigate the erosive effects of ice along the Susitana River are coarse-paved bed material, bank-attached ice, vegetation rootmats, and channel form and local hydraulics [11]. Two ice-related factors limiting bank erosion are frozen banks, which limits the removal of sediments, and shear walls, which protect the bank from contact with moving ice [10].

Moving ice can erode riverbanks by grinding the bank surface, by transporting boulders, and undercutting bank materials [7,10,12–14]. The extent of gouging by impacting ice is less if the top of the bank has larger vegetation with protective rootmats rather vegetation with shallow rooting plants. However, ice rubble moving along the bank can shear rootmats, leaving a near-vertical bank face exposed to erosion by flows at stages above the gravel bank toe [10]. Furthermore, the effects of flow-driven rubble-ice are of greater severity where ice rubble directly impacts a bank, such as at the head of an island or bar, compared to other locations, such as the side of mid-channel islands [10].

River banks are an important source of sediment in natural streams. The annual thermal cycle is found to affect this source greatly by affecting the stability of the stream banks [15]. Gatto [8] identifies several bank erosion mechanisms that can contribute to increases in suspended sediment during the breakup period. Turcotte et al. [13] also reviews the potential effects of streambank and ice processes on sediment supply.

The occurrence and magnitude of bed erosion (i.e., scour) depends upon several factors, such as channel slope and planform, the quantity and hydraulics of channel flow (which can be affected by an ice cover), and the properties of the bed material. For non-cohesive sediment, the properties of bed material affecting scour are particle size, shape, density, and position. For cohesive bed material, the strength of the cohesive bond between the particles affects erosion [16].

The capacity of the flow to erode channel boundaries and transport sediment is related to bed shear [17]. The average bed shear stress, τ_b, is customarily expressed as:

$$\tau_b = \gamma \, R_b \, S_f \tag{1}$$

where γ = unit weight of water = 9810 N/m^3; R_b = bed-controlled hydraulic radius $\approx$ average water depth under open-water flow conditions in most rivers; and S_f = friction slope. Taking into account the Manning resistance equation, it can be further shown that:

$$\tau_b = \gamma \, n_b^{3/2} \, S_f^{1/4} \, U^{3/2} \tag{2}$$

where n_b = the Manning's resistance coefficient for the channel bed, and U = average flow velocity in m/s. Equation (2) is particularly helpful during ice runs because the surface velocity can be estimated by timing moving ice floes and the result multiplied by ~0.9 (per Equation (6), Section 5) to obtain average flow velocity. Noting the small exponent of the friction slope in Equation (2), reasonable estimates of the shear stress can be obtained by substituting the water-surface slope in its place. Bed shear stress can be especially high during the passage of javes because both U and S_f are considerably greater than their pre-jave values. Beltaos et al. [18] estimates a bed shear stress of approximately 120 Pa during the passage of a major jave on the lower Athabasca River, a value that was corroborated by more detailed analysis of jave characteristics [19]. By comparison, the highest known discharge in the same reach, which occurred under open-water conditions, generated an estimated bed shear stress of only 35 Pa. Erosional capacity has also been expressed in terms of unit stream power, which is defined as the product τU. Annandale [20] presents an erodibility index method based on the concept of a critical value of unit stream power that can be used to compute the hydraulic conditions under which erosion will be initiated in a wide range of materials.

2.2. Sediment Transport

Sediment is transported due to the combined effects of water flow and gravitational forces acting on the sediment. Transport modes differ depending upon sediment size, shape, and density, with larger particles sliding or bouncing along the riverbed, and finer particles suspended in the flowing water. Sediment can be transported in solution (dissolved load), in suspension (suspended load), or along the bottom of the river (bed load). Sediments transported as bed load and as suspended load have different particle size distributions, with suspended-sediment particles usually being considerably smaller than the bed-load particles. A change in land use, for example, from forest to agricultural (or logging), can substantially modify sediment transport modes and intensities in river systems by increasing fine sediment supply [21]. Many sediment transport formulae exist, but some are for total load, some are for suspended load, and some are for bed load. No formula applies to all flow and sediment conditions. Different formulae yield different results, and thus, the accuracy of results depends upon how well the formula applies to a river's flow and sediment regime.

In gravel-bed rivers, bed load will be a major portion of total sediment load. Basically, it is possible to compute the suspended-load capacity based upon the type of bed material and the flow conditions.

In-stream sediment transport depends on the sediment supply and the sediment transport capacity [13]. Sources of suspended sediment during breakup include particulate material deposited by wind on the ice cover prior to breakup, sediment carried by runoff from land surfaces, material removed from the banks by freeze-thaw cycles, and the bed and bank soils eroded, abraded, or gouged by moving ice and water. A portion of the suspended sediment in a river (such as clays) may be so fine that the particles do not rely on turbulence but are kept in suspension by thermal molecular agitation (Brownian motion) and are referred to as wash load, as it washes through the system [22]. If the sediment supply is limited, the actual sediment loads will be less that the sediment transport capacity of the stream. The concentration of suspended sediment is generally several orders of magnitude below the channel's capacity for sediment transport [23]. Therefore, the rate of supply is the dominant control on suspended-sediment concentration. Considerable scatter is evident when the concentration of suspended sediment is plotted against discharge. Temporal changes in suspended-sediment concentrations with hydroclimatic conditions may create a hysteresis loop because the rate of fine sediment supplied to the flow is greater during the rising limb of the hydrograph compared to the falling limb [24]. Sediment deposited and stored on the channel bed between storms is entrained by the increasing velocities during the rising limb, leaving less sediment supplied to the flow during the falling limb [23].

The suspended load transport per unit width is computed by integration over the flow depth of suspended-sediment concentration times the local mean longitudinal flow velocity. Integration of this load across the entire channel width results in the total suspended load, which often dominates total sediment flux, and is thus a major contributor to sediment influx to lakes and oceans and to deposition onto flood plains.

The particle fall velocity and the sediment diffusion coefficient are the main controlling hydraulic parameters for the suspended load [25]. Initiation of sediment suspension from the riverbed begins when the turbulent eddies have dominant vertical velocity components that exceed the particle fall velocity. The maximum value of the vertical turbulence intensity is of the same order as the bed-shear velocity [25]. The particle fall velocity is a function primarily of the kinematic fluid viscosity, sediment-specific gravity, and the representative particle diameter of suspended sediment. Different equations for particle fall velocity have been proposed for different ranges of sediment size. The sediment diffusion coefficient governs the concentration profile of suspended sediment [26]. Temperature variations impact rivers that transport a preponderance of fine sediment, a characteristic of many large rivers [27]. River temperature controls fluid density and viscosity and hence sediment transport.

Total suspended solids (TSS) and turbidity are often used as indicators of suspended sediment. TSS is a direct measurement of quantity of solid material, both organic and inorganic, per volume of water. Although SSC (suspended sediment concentration) and TSS are both measurements of the solid-phase material within a water column, values of TSS and SSC can differ due to different laboratory procedures used in their determination. Turbidity is a measurement of the passage of light though a liquid, which often can be measured or qualitatively observed onsite. Turbidity can be affected by solid and dissolved substances. Mean turbidities of ice-covered river water of the Kokemäenjoki River, Finland, were observed under similar discharges (<300 m^3/s) to be 1.5 to 3.3 times less than during summer or winter ice-free conditions [6]. TSS can be used to calculate sedimentation rates, while it is generally considered that turbidity cannot (e.g., [28]). In general, SSC, TSS, and turbidity data collected from natural water are not directly comparable and should not be used interchangeably.

2.3. Sediment Transport during an Ice Season

Lower streamflow during winter periods occurs due to limited surface runoff during cold weather. Mid-winter flow velocities are relatively low under a stable ice cover, thereby lowering bed shear and associated sediment transport. TSS and shear stress analyses of the Kokemäenjoki River, Finland, indicate that the sediment transportation in the river declines under ice cover [6].

Anchor-ice-rafting can have a major role in moving bed sediment, as anchor ice released from the bed can transport some bed sediment to which the ice was attached. Based on field observations and measurements in the Laramie River, USA, Kempema et al. [29] found anchor-ice-rafted sediment is coarser than fluvially transported bedload sediment; that is, ice rafting (not peak spring flows) transported the largest particles. The amount of sediment transport by anchor ice could be significant in some regulated rivers where open water persists throughout the winter season, letting anchor ice form and release often [30]. For three regulated rivers in Alberta, Canada (the North Saskatchewan, Peace, and Kananaskis rivers), the average sediment concentration contained in released, floating anchor ice was found to be 28.2 g/L with a standard deviation of 33.2 g/L and a median of 18.4 g/L [30].

The ice cover itself can also contain sediment during the ice season. High flows at freezeup coupled with irregular winter flows are observed to result in sediment embedded in the ice of both the Athabasca and Peace River systems [31]. Ice-entrained sediment can be transported downstream with broken ice during breakup and enter the water column as the ice melts.

For the Tanana River near Fairbanks, Alaska, Lawson et al. [32] reported a lower ratio of suspended-sediment load to bed load during the winter (February–March 1984) compared to the summer. Measured SSC values during the period of ice cover were 51 mg/L to 152 mg/L, considerably higher than the winter pre-breakup SSCs reported by [33,34]. Possible reasons for this are finer bed and bank material and increased velocities in slush-confined channels under the solid ice cover. This shows that suspended-sediment concentration is dependent upon local conditions.

Using a numerical model to evaluate the effects of various parameters on shear stresses during ice motion, Ferrick and Weyrick [35] found the amplitude of increase in suspended-sediment-transport capacity increased with ice velocity and acceleration and decreased with relative ice roughness. Higher initial energy gradients led to smaller increases in sediment transport capacities than in flow velocity [35].

3. Overview of Ice Breakup Processes

3.1. Physical Processes

The breakup is typically a brief event spanning the transition from the relatively complete and integral winter ice cover to the open-water condition. Often attended by rapid increases in river stage and flow velocities, the breakup of river ice can disrupt ecosystems, abrade stream banks, alter channel morphology, scour riverbeds, and transport large quantities of sediment and associated contaminants, and modify water quality. Detailed background information on breakup processes and their socio-economic and environmental impacts can be found in the book *River Ice Breakup* [2].

Ice breakup is triggered by mild weather, which can generate snowmelt and rainfall on the one hand and, on the other hand, thermally degrade the stationary winter ice over. As river flow increases, so do the hydrodynamic forces that are applied on the ice cover, while the rising water levels reduce ice attachment to the river banks and bed. At the same time, the ice cover becomes more susceptible to dislodgment and fracture via thermally induced reductions in thickness and strength. Eventually, large segments of the fragmented cover are dislodged and mobilized by the flow. This is the onset of breakup and is followed by the drive, i.e., the transport and further breakdown of large ice sheets into smaller blocks and rubble. The onset is governed by many factors, including channel morphology, which is highly variable along the river. It is thus common to find reaches where breakup has started alternating with reaches where the winter ice cover has not yet moved.

Invariably, this situation leads to jamming because ice blocks moving down the river in one reach encounter stationary ice cover in another reach and begin to pile up behind it, initiating a jam, which can be tens of metres or, more often, several to many kilometres long. Ice-jam residence times vary from a few minutes to many days depending on a variety of factors, such as local channel morphology and bathymetry, variation of discharge (increasing, steady, decreasing), and rate of thermal degradation of the intact ice cover downstream of the jam (air temperature, precipitation, solar radiation, cloudiness). The javethat is generated by an abrupt release of an ice jam can dislodge and break up long sections of intact downstream ice. On occasion, the stationary ice is too strong or the jave too attenuated, and the ice run that lags behind the jave is arrested, forming a new jam. In this manner, more and more ice is broken up and carried down the river until the final jam releases. This is the start of the wash or final clearance of ice from a river reach, though grounded remnants of ice rubble and ice slabs may linger for weeks.

In the colder continental parts of Canada, such as the Prairies or Territories, we are most familiar with a single event, the spring breakup, which is triggered by snowmelt. In more temperate regions, however, such as parts of Atlantic Canada, Quebec, Ontario, and British Columbia, events called mid-winter thaws are common. Usually occurring in January or February, they consist of a few days of mild weather and often come with significant rainfall. River flows may rise very rapidly and sufficiently to trigger breakup on many local rivers. This is the mid-winter breakup, which can be more severe than a spring event because of the sharp rise in flow that results from the rain-snowmelt combination. Moreover, dealing with the aftermath of flooding is hampered by the cold weather that resumes in a few days, while many mid-winter jams do not release but freeze in place, posing an additional threat during subsequent runoff events.

Ice jams and javes are the primary breakup phenomena that contribute to erosion, deposition, and transport of sediment. Owing to their large aggregate thickness and underside roughness, major ice jams raise water levels far above what would suffice to pass the prevailing flow under open water conditions. This can lead to flooding and deposition of large quantities of sediment on river floodplains. Typically, the longitudinal profile of an ice jam (Figure 1) entails the potential for bed scour, primarily near the toe, where significantly increased velocities and shear stresses are encountered owing to the reduced under-ice flow area. It is evident in Figure 1 that as the flow approaches the toe of the jam, the associated bed shear stress will develop a positive longitudinal gradient followed by a negative gradient farther downstream. If the magnitude of the shear stress is sufficient to erode the bed, scour will occur upstream and deposition downstream of the toe location. To a lesser degree, the same may occur immediately downstream of the head of the jam, where a positive shear stress gradient is also likely to develop. Upstream of the head, the relatively large water depth is associated with smaller flow velocities and shear stresses, which decrease in the downstream direction. Here, sediment deposition will likely occur depending on the magnitude and particle size of any incoming sediment, be it suspended or bed load.

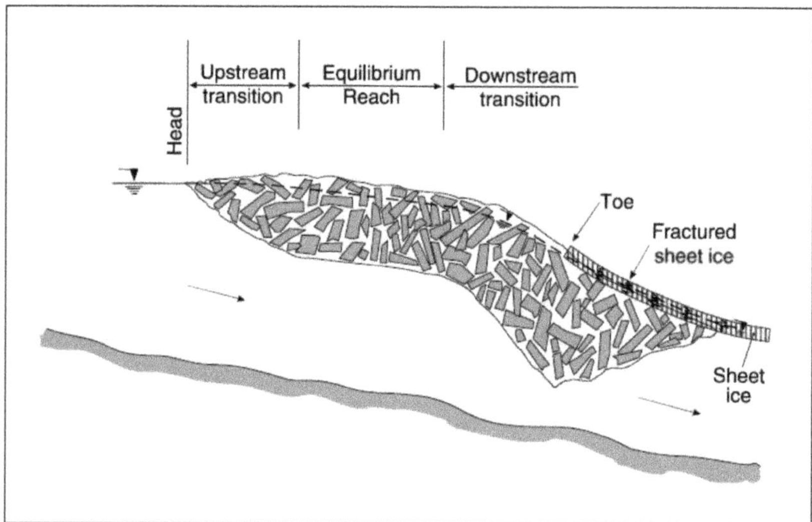

Figure 1. Schematic illustration of a fully developed ice jam. The equilibrium reach is usually much longer than indicated in the sketch but may be absent in short jams. Reproduced from [36]. Crown copyright, published by CGU-HS Committee on River Ice Processes and the Environment.

The blockage and large hydraulic resistance to flow created by ice jams can divert flow away from the main channel and into secondary channels. As a result, local flow velocities may increase far beyond what is expected under non-jammed conditions. Gerard [37] observed ~3 m high standing waves in a side channel of the Athabasca River at Fort McMurray following partial failure of an ice jam on 22 April 1974. From crude estimates of the wave length, he estimated the flow velocity to have been between 4 and 7 m/s. The partial failure had "concentrated the total release into a small side channel near the left bank".

Typically, ice jams release abruptly as a result of increasing river flow and/or weakening of the downstream stationary sheet-ice cover. If the toe of the jam happens to be grounded, ice blocks in contact with the channel boundary could gouge the riverbed and banks, changing local morphology and releasing large amounts of sediment into the flow [38]. Moreover, a release generates a positive and a negative wave, respectively, moving in the downstream and upstream directions. The rapid decrease in water levels upstream of the toe of the jam can cause extensive bank failure due to pore pressure imbalance. However, it is the sharp positive wave, or jave, that is much more dynamic than a runoff wave and can cause the bulk of sediment erosion and transport. Early work on javes was discussed in the comprehensive review and synthesis by [39]. More recent research further elucidated jave characteristics, including but not limited to [3,40–43].

Examples of jave waveforms are illustrated in Figure 2. The rise to the peak stage lasted 15 min in total, but the first 5 min brought about a rise of ~4 m at the farthest upstream station [41]. This extreme rate of rise is one of the highest known to date. Comparable rates of rise have also been recorded in the Restigouche River (~3 m in 6 min; [44]) and the Athabasca River (~4.5 m in 6 min; [18]).

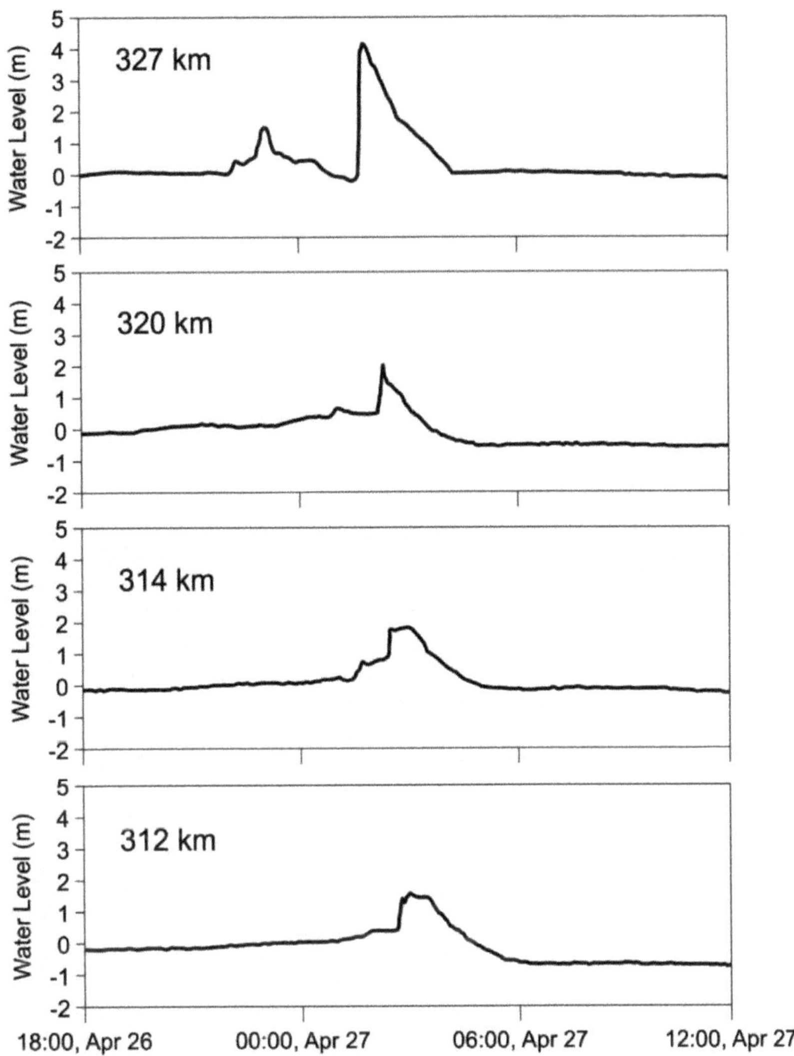

Figure 2. Passage of a 2002 jave in a 15 km reach of the Athabasca River, captured by temporary water level gauges at different locations above Fort McMurray, Alberta, Canada. The km values represent approximate river distances upstream of the Athabasca River mouth; the Grant MacEwan bridge at Fort McMurray s located at km ~295. Adapted from [41], with changes.

Theoretical considerations suggest that the leading-edge celerity of a jave should not exceed the celerity of a gravity wave, C_g, or be less than the celerity of a kinematic wave, C_k, [3,43,45]. The theoretical celerities are defined as:

$$C_g \approx U_o + \sqrt{gh_o} \tag{3}$$

$$C_k \approx \beta\, U_o \tag{4}$$

in which U_o and h_o = pre-jave mean flow velocity and depth, respectively; g = gravitational acceleration; and β = coefficient equal to 3/2 or 5/3 depending on whether the Chezy or the Manning equation is used to relate velocity to hydraulic radius and friction slope.

Observations on the Hay River, NWT, Canada, showed that the leading-edge celerity remained close to C_g for at least 1.1 to 2.9 ice-jam lengths of travel, with the crest celerity being a fraction of C_g [43]. Beyond this distance, both celerities decreased, with the crest celerity approaching C_k.

The presence of moving ice renders problematic the measurement of velocity and discharge. However, such hydrodynamic variables, including the bed shear stress, can be inferred using the Rising Limb Analysis Method (RLAM), developed by [46]. Application of the RLAM to numerous recorded waveforms has revealed that javes can greatly amplify hydrodynamic variables especially where the wave height is large and the rising limb brief, as illustrated in Figure 3. The scatter is due to partial only accounting of the several independent variables arising from dimensional analysis; therefore, Figure 3 provides estimates of magnitude range rather than a single value of the amplification ratio. For a moderately sharp jave, rising 3 m in 30 min, with initial flow depth and velocity of 3 m and 1 m/s, respectively, the value of the abscissa works out to be 0.26, indicating a four- to six-fold amplification. For waves propagating under a stationary ice cover that is dislodged prior to the wave crest, the quantity τ_{jave} may be less than the peak shear stress because the open-water flow condition can generate higher shear than what RLAM calculates, assuming stationary cover throughout the duration of the rising limb [19]. Therefore, the amplification ratios indicated by the solid triangles in Figure 3 may be equal to or greater than the actual peak values generated by the respective javes.

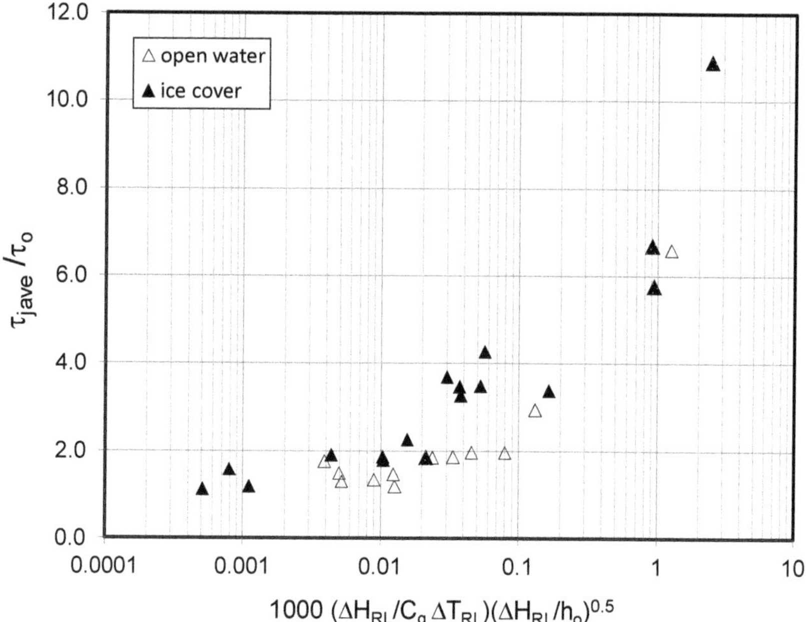

Figure 3. Shear stress amplification ratio versus a product-combination of independent dimensionless variables for javes propagating in open-water and in ice-covered reaches. Symbols: τ_{jave}, approximate peak bed shear stress; τ_o, pre-jave shear stress; ΔH_{RL}, wave height; ΔT_{RL}, duration of rising limb. Adapted from [3]. Crown copyright, published by Elsevier.

Beltaos et al. [47] and Beltaos [19] reported bed shear stresses of up to 40 and 140 Pa generated by javes in the Saint John and Athabasca Rivers, respectively. The higher value occurred very near the toe of the released jam and likely represents the highest that occurred along the river during the passage of that jave. Though the corresponding breakup events were of the dynamic kind, neither one was extreme.

High velocities and shear stresses suggest potential for intense bed and bank erosion as well as transport of large quantities of sediment both as bed and as suspended load. Visual evidence indicates that the water takes on a deep muddy hue when a river opens in dynamic fashion, as in the case of a jave. River conditions near the Fort McMurray golf course were characterized by a deep muddy hue shortly after the arrival of a jave and dislodgment of the winter ice cover (Figure 4). At this early phase of the ice run, most of the river surface is covered by ice plates and rubble. However, the surface of the water and various churning ice blocks in it can be seen in the lower left corner of the photo; its colour suggests a very high concentration of suspended sediment. Over time, concentrations gradually decrease, but elevated values linger (Figure 5).

Figure 4. Early phase of ice run near Fort McMurray Golf Course, 27 April 2014. Photo taken by time-lapse camera installed on the left (north) river bank and looking across and slightly downstream. The water surface is exposed in the lower left portion of the image. Local river width ~400 m. From [18], Crown copyright, published by Elsevier.

Numerous field observations and related data analysis (e.g., [3,43,44]) have conclusively shown that the rising limb of the jave advances at a greater rate than the ice run, which comprises rubble from the released jam. Typically, a jave advances at first under stationary ice cover, which is often mobilized at some point along the rising limb; large ice sheets then pass by for a relatively brief time, followed by bank-to-bank ice rubble. This sequence has additional implications to erosional processes, namely:

(i)   Swiftly moving large ice sheets that follow dislodgment of the stationary ice cover can scrape and gouge riverbanks over extended channel distances.

(ii) Heavy runs of ice rubble that follow the large ice sheets often result in formation of thick grounded strips along the river banks. Such strips shield the banks from further erosion either by ice abrasion and gouging or by the high-velocity water underneath the moving rubble.

Figure 5. Appearance of water surface after an ice run near Mountain Rapids, ~12 km upstream of Grant MacEwan bridge at Fort McMurray, Alberta, 1 May 2013. As the jave progresses downstream, concentrations of suspended sediment in upstream reaches gradually decrease but remain high relative to typical open-water conditions. Photo by S. Beltaos.

3.2. Sediment Transport Modelling in Ice-Laden Rivers

Numerous laboratory experiments in channels with stationary covers of constant thickness have shown that bed sediment transport equations previously developed for open-channel flow can be suitably modified for under-ice transport [12]. Knack and Shen [48,49] further developed such equations and incorporated them in a hydrodynamic model, which successfully simulated laboratory scour and deposition experiments under conditions of steady incoming flow.

Bed sediment transport under the highly dynamic flow conditions that prevail upon and after release of ice jams is examined by coupling a non-hydrostatic 3D hydrodynamic model with bed erosion and deposition equations to track the changing bathymetry of a riverbed [50]. Initially, an ice jam was modelled as a rigid body of water near the free surface that constricts the flow. This "jam" did not exchange mass or momentum with the stream, but the "ice" body could have a realistic shape and offer resistance to the flow of water through the constriction. The release was modeled by suddenly enabling the "ice" to flow and exchange mass and momentum with the water. This model was applied to the extreme 1984 ice jam event in the St. Clair River to determine whether bed scour could have occurred. The simulations suggested that events like the 1984 ice jam can cause significant scouring of the St. Clair River bed upon their release, and the authors recommended future field research to ascertain this result.

Dynamic changes in the state of the pre-release stationary ice cover, which is typically located downstream of an ice jam and is often mobilized and broken up by advancing javes over long distances, add complexity to numerical simulation. The Knack and Shen model [48,49] can simulate flow hydrodynamics under such conditions and hence compute bed sediment transport, but this capability has not yet been tested, owing to lack of relevant data (Dr. I. Knack, pers. comm., March 2021, [51]).

Field observations and data [52,53] suggest that a large component of the sediment load being delivered during breakup derives from bank erosion processes. Although bank erosion rates have been quantified by numerous investigators for open-water flow conditions [54], this knowledge has not yet been incorporated in hydrodynamic river-ice models.

4. Suspended-Sediment Concentrations Shortly before and during the Breakup Period

Measurements of SSCs and sediment characteristics during the pre-breakup and breakup periods are discussed in this section, based on a review of literature. During the breakup of river ice covers, rising discharge and moving ice increase the potential for erosion due to rising discharge and moving ice. The bulk of the sediment load is thus delivered on the rising limb of the flow hydrograph, evincing constraints on the amount of sediment supply.

Prowse [33] reported measurements and analyses of SSC during the 1987 pre-breakup and breakup periods on the Liard River near its confluence with the Mackenzie River. All sampling was done 300–500 m from the left bank in an approximately 700-m wide reach used by the Water Survey of Canada to monitor discharge and sediment of the lower Liard River (hydrometric station No. 10ED002, Liard River near the Mouth, drainage area = 275,000 km^2). Up until 26 April, SSCs remained below 10 mg/L but then began to steadily rise as breakup occurred upstream and along tributaries. Measurements made on 1 May from a helicopter during the ice fracturing and fragmenting period revealed a sediment concentration of 70 mg/L. The general rise in SSC prior to breakup at the sampling site is probably attributable to a rise in velocities and discharge and additional sediment contributions from upstream tributaries, including some more southern rivers which break up earlier [33]. Ice cleared the reach on 2 May, and a local, small jam occurred downstream, increasing water levels in the sampling reach. A sample on 4 May revealed that the sediment concentration had risen to approximately 120 mg/L. The highest measured SSC during the 1987 breakup of approximately 1067 mg/L occurred on 6 May during release of jammed ice remaining on the lower Liard [33]. This value far exceeded normal sediment concentrations for open-water conditions with equivalent discharge and was closer to annual peak-sediment concentrations recorded at two to five times greater discharge. Upstream ice-induced erosion is most likely responsible for the rapid increase during the 4–6 May breakup period [33]. Higher sediment concentrations could be produced by more dynamic breakup events than the moderate 1987 event, which was characterized by a nonuniform downstream breakup progression typical of a thermal breakup.

Milburn and Prowse [55] presents data from a 1993 sampling program on the Liard River and further discussed the results of suspended-sediment sampling conducted in 1987. During both the 1987 and 1993 data gathering, samplers were lowered through holes in the ice during pre-breakup and from a helicopter positioned over leads during active breakup. The 1987 breakup stage and the 1993 breakup stage were just below and above the maximum open-water stage, respectively, and are more representative of average than extreme breakup severity. In both years, suspended-sediment concentrations remained below 10 mg/L two weeks before breakup but then rose by an order of magnitude prior to breakup. This increase in pre-breakup suspended sediment could be due to resuspension of fine-grained sediment deposited during the low-flow winter period [55]. As reported in [33,56], SSCs in 1987 rose more as breakup progressed, reaching a maximum of 1067 mg/L with a concurrent mean daily discharge of 2550 m^3/s. The 1987 high SSCs could be related to an upstream release of an ice jam. In 1993, SSCs rose to 291 mg/L (at 2280 m^3/s) just before breakup and peaked at 331 mg/L (at 2480 m^3/s) during the final ice run [55].

The average proportions of clay, silt, and fine sands in the five samples collected in 1987 just before and at breakup are 18, 66, and 16%, respectively; they are much finer than that when measured during the open-water period [55]. The high percentage of very fine material in these samples supports remobilization of over-winter accumulated sediment material as a source of pre- and early breakup suspended sediment. If particle-size analysis had continued later during breakup, it is expected that later samples would show a shift toward higher percentages of fine sand fractions because of increased erosive action [55].

Measured sediment concentration in the Hequ Reach of the Yellow River, China, increases with flow discharge and during ice breakup, while the grain size distribution of sediment can show seasonal variations [57]. The suspended load carried by the flow during the winter is coarser than that during formation of ice jams and the breakup period, as the supply of finer sediment from land sources is reduced, and some finer sediment is entrained in ice, therefore making the coarser material of the riverbed the predominant source of sediment [57].

Toniolo et al. [58] reports on field measurements carried out during the spring breakup of 2011 on seven pristine streams located in the National Petroleum Reserve in Alaska. Estimates of the Shields parameter indicated finer sediments could be moved in suspension but that coarse sand would move as bedload. The highest SSC of 186.55 mg/L reported in the paper occurred on 2 June on the Fish River, and the corresponding suspended-sediment load was estimated to be nearly 14.9 kg/s. Suspended-sediment concentrations more than 125 mg/L were also reported for the Seebee Creek and Prince Creek. In general, changes in suspended-sediment concentrations can be associated with variations in discharge and with moving ice.

Moore et al. [59] presents hourly and daily-averaged SSC values on the Nelson River, Manitoba, estimated from acoustic data during hydropeaking operations at the Limestone Generating Station. Daily-averaged SSCs increase from around 7 mg/L to 130 mg/L during the breakup period. Suspended-sediment concentration is observed at breakup to reach 10 times SSC values under a stable ice cover [59]. Hourly average peak values of SSC were much higher, attaining ~550 mg/L on one occasion (Figure 6).

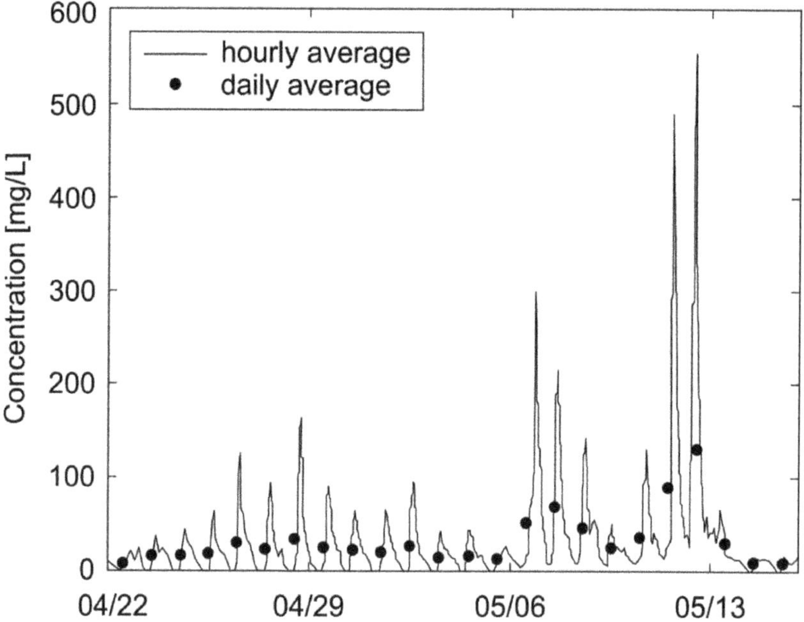

Figure 6. Concentration estimated from 2012 acoustic data at 2.24 m from the bed at a Nelson River site located 50 km downstream of the Limestone Generating Station. Adapted from [59], with authors' permission.

Environment Canada and the New Brunswick Department of the Environment (NB-DOE) undertook a joint five-year study (1993–1997) of ice breakup and jamming along the Saint John River (SJR) from the Dickey area in Maine, United States, to St. Leonard, New Brunswick, Canada. Sediment-related aspects of the study, including methodology and findings, are presented in Beltaos and Burrell [34,52]. The main findings are summarized in the next few paragraphs.

The temporal and spatial variation of the SSC was investigated by acquiring and analysing suspended-sediment samples at four bridges along the Saint John River from Dickey, Maine, USA to Saint-Leonard, New Brunswick, Canada [34,52]. Using the bridges as platforms, samples were acquired by dipping standard 750-mL glass bottles into the river or inserting them in a P72 point-integrating fish-type sampler. During typical freezeup conditions, measured SSCs along the Saint John River ranged from 0 to ~2 mg/L [34]. Concentrations increased on occasion, when the freezeup process was interrupted by small runoff from minor rain-on-snow events, decreasing again as the flow subsided. Similar values of SSC (~0 to 2 mg/L) were also found during steady-flow conditions in the winter right up to late March and even in the first half of April [34]. Peak runoff-induced sediment concentrations during the 1993- to 1997-inclusive breakup events ranged from ~35 to 150 mg/L (Figure 7), much greater than under open-water conditions. As described in Section 4, even higher SSCs were measured during brief sediment "pulses" accompanying the passage of javes. The suspended-sediment loads associated with breakup events can be the greatest fraction of the total annual loads [34].

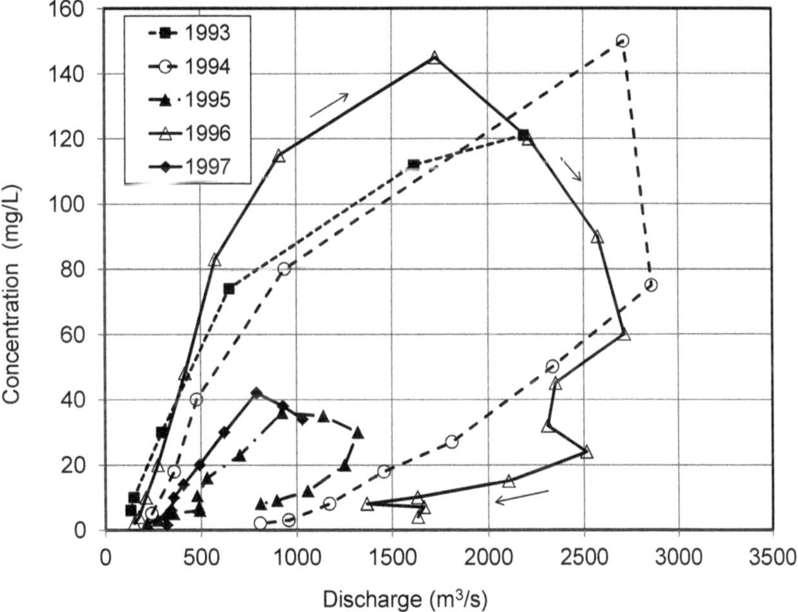

Figure 7. Daily mean values of runoff-generated SSC versus daily mean discharge during the spring freshet. Saint John River at Clair, 1993 to 1997. Time increases in the direction indicated by the arrows. From [34], Crown copyright, published by Elsevier.

The SSCs gradually decrease as river flow decreases and water levels subside; however, SSCs were found to decrease to pre-event values much faster than does discharge, with the peak discharge lagging 1–3 days later than the peak SSC, thereby indicating a limit on available sediment supply during the breakup event. In many rivers, such as the Saint John River, the rate of sediment supply rather than the flow capacity limits the amount of sediment transport. The hysteresis loop in the sediment rating curve of SSC versus discharge (Figure 7) is an indication of limits to sediment supply. The relationship between suspended-sediment loads and flow discharge can also vary depending upon the type of hydroclimatic event and its intensity. Though concentration–discharge relationships involve large scatter, sediment loads delivered during specific runoff events correlated well with corresponding water volumes [3,4].

Using a laboratory Malvern laser particle-size analyser unit, many dip samples obtained during the 1993 breakup along the Saint John River were analysed to determine primary particle sizes during the actual breakup event. Each sample was first hand shaken and analysed and then sonicated to destroy any flocs that might have been present and analysed again [52]. Results indicate that primary sediment particles in suspension are orders of magnitude finer ($D_{50} \sim 10$ μm on average) than the river bed material, suggesting that the bulk of the suspended sediment does not originate from the bed of the river. The strongly looped appearance of C vs. Q graphs (Figure 7) supports this finding.

To investigate whether flocculation occurs in the river environment, a field Malvern unit was deployed at two bridge sites (Edmundston, St. Leonard) on 24 August 1993 (flow $\sim$75 m^3/s) and at four bridge sites (Dickey, Clair, Edmundston, and St. Leonard) in May 1997 (flow $\sim$1500 m^3/s). During the August 1993 measurements, D_{10}, D_{50}, and D_{90} showed little lateral variability with average values of $\sim$4, 14, and 35 μm, respectively, while the May 1997 results indicated greater variability along and across the channel ($D_{50} \sim$11 to 29 μm). Comparison with corresponding sonicated samples indicated significant flocculation at the higher flow condition (Figure 8). The presence of moving ice precluded deployment of the field Malvern unit during breakup conditions. Indirect evidence suggests that flocculation may be even more intense during breakup [52].

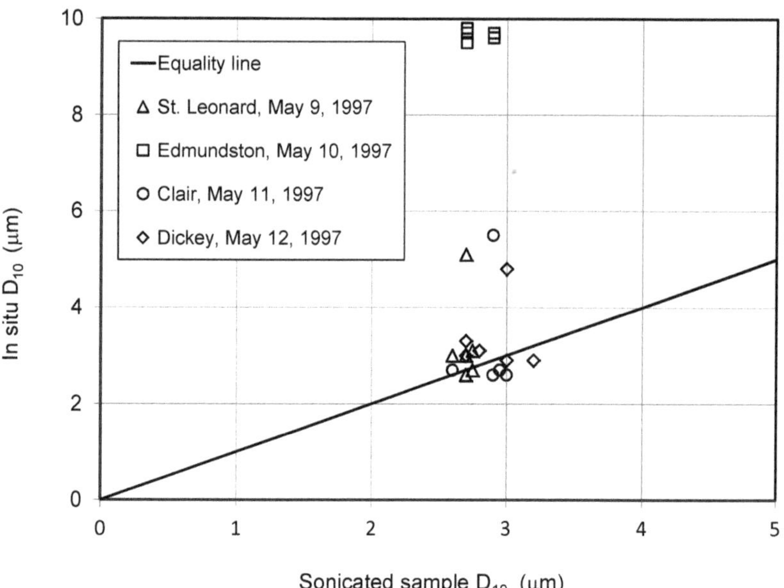

Figure 8. Comparing in situ sediment size to primary particle size: ninetieth percentile. From Beltaos and Burrell (2016b). Crown copyright, published by Elsevier.

For the Lower Athabasca River, a positive correlation was found between SSC and flow discharge using sediment data for the WSC (Water Survey of Canada) hydrometric gauge located a few km downstream of Fort McMurray [60]. The WSC data set, which consists of published daily values of sediment load during the years 1969–1972, exhibits large scatter, which can be largely enveloped by the relationships [19]:

$$Load = 7 \times 10^{-6} Q^3 \quad \text{to} \quad 5 \times 10^{-5} Q^3 \tag{5}$$

in which the load is expressed in tonnes per day, and Q = flow discharge in m^3/s. Since the load equals $Q \times SSC$, it can be shown via Equation (5) that SSC (in mg/L) would range from $8 \times 10^{-5} \times Q^2$ to $7 \times 10^{-4} \times Q^2$. For typical breakup flows at this location (~500 to 1000 m^3/s), SSC values of 20 to 700 mg/L should be common; they can be much larger during the passage of javes, when discharge can be several times the pre-jave value.

Skolasińska and Nowak [61] analysed data collected at the Sieradz gauging station (hydrometric basin area =8156 km^2) in Central Poland during the period of 1961–1980 and found that changes in SSC on the Warta River occurred during periods of river ice. Suspended-sediment concentrations increased gradually during freezeup and increased rapidly when the ice cover broke, during which time ice movement caused lateral erosion of the river banks [61]. From the data presented in the paper, the suspended-sediment concentration seems to have exceeded 70 mg/L during the breakup period.

5. Javes and Sediment Pulses

As discussed in Section 3.1, the highly dynamic jave is a wave form that occurs in ice-forming rivers and can cause significant bed scour and bank erosion. Examples of jave observations are presented in this section along with observed jave celerities and concomitant SSCs. Observations and data on a jave that occurred on 26 April 2002 on the Athabasca River were reported in [62]. Upon jam release, the water level just downstream was observed to rise 4.4 m in 15 min as the jave passed, and jave celerity in the reach was estimated to be 4.05 m/s, decreasing as it moved downstream). The peak ice concentration was observed to lag the crest of the wave as it travelled downstream [62]. Beltaos and Burrell [44] reports on field observations and measurements of three javes on different New Brunswick rivers. The passage of the jave created upon the release on 13 April 2002 of an ice jam in the Saint John River (toe 7.8 km upstream of Clair) was evident in the output of a pressure logger, installed 900 m downstream of the toe and at two downstream hydrometric stations (Fort Kent and Edmundston). An inspection of the data revealed that the celerities of the jave leading edge, peak stage, and trailing edge differed [44]. Between the surge meter and the Fort Kent gauge (distance of 8.2 km), the celerity of the leading edge and the celerity of the jave peak were 5.5 m/s and 2.5 m/s, respectively. For the stretch of river from Fort Kent to Edmundston (distance of 34 km), the celerities of the leading edge, peak, and trailing edge were 3.5 m/s, 2.1 m/s, and 1.7 m/s, respectively [44]. A review of historic records of several ice-jam release events on the Athabasca River at Fort McMurray, Alberta, indicates that release waves more than 3 m in height and propagation speeds in excess of 4 to 5 m/s were common [41]. Factors affecting jave propagation speeds were the interaction of water and ice within the jam, the size of the wave upon release, the distance travelled, channel geometry, downstream ice conditions, and the temporally jamming and re-release of released ice [41]. Detailed documentations of several large javes in the Hay and Athabasca Rivers are presented in [43] and [18], respectively.

The bed shear stress, which quantifies the erosive power of the flow, was deduced from measured waveforms using the RLAM and shown to be considerably amplified during the passage of a jave. Peak values of ~40, 100, and 140 Pa were reported for javes in the Saint John, Restigouche, and Athabasca Rivers, respectively [3,19]. Erosive power is also manifested in water velocity, which can be visually estimated by timing various ice floes in ice runs. Ice velocities of 3 m/s are common [3]. Ice run speeds ranging from 2.5 to 4.2 m/s during the 2007 breakup event on the Athabasca River above Fort McMurray were reported in [42]. Notably, extreme ice speeds have been reported in this area over the years, the highest being 7.4 m/s [63]. Ice speeds of ~5 m/s in the same area and loss of two pressure loggers as a result of entire bank segments being carried away by javes were reported in [18].

With reference to ice runs, it is noted that ice speed reflects the velocity of the near-surface layer of the flow, while the mean flow velocity in the same vertical would be somewhat less; according to the well-known logarithmic distribution, the mean-to-surface velocity ratio is given by:

$$r_U = \frac{U_{mean}}{U_{surface}} = \frac{1}{1 + 7.82\ n_b h^{-1/6}} \quad \text{(metric units)} \tag{6}$$

in which U = velocity, n_b = bed Manning coefficient, and h = flow depth [64]. For likely ranges of 0.02 to 0.03 for n_b and 2 to 10 m for h, r_U ranges from 0.83 (0.03, 2 m) to 0.90 (0.02, 10 m). For quick estimates, the midpoint of this range (0.87) is a reasonable choice. For instance, a surface velocity of 7.4 m/s translates to $U_{mean} \sim 6.4$ m/s. In turn, Equation (2) (Section 2) implies a bed shear stress of ~150 Pa even with the steady-state slope (~0.001) of the applicable reach. It is very likely that such high velocities occur on the rising limb of a jave, i.e., when the water surface slope and friction slope are considerably higher than the steady-state value. Frontal jave slopes of ~0.003 are known to occur in the Athabasca River above Fort McMurray (Figure 16 of [18]).

During breakup, sediment pulses can occur as brief but sharply peaking concentrations of suspended sediment. A pulse was most likely responsible for the SSC value of 1067 mg/L that was measured in the Liard River on 6 May 1987, which was an order of magnitude higher than previously measured values [33]. Sediment pulses can be deleterious to aquatic life and contribute significantly to annual sediment loads [34].

Comprehensive datasets on the height and duration of suspended-sediment pulses occurring during five breakup events (1993–1997) on the upper Saint John River (Maine (USA) and New Brunswick (Canada)) are presented in [53]. Figure 9 combines sedigraphs of the 1994 breakup event at four bridge sites along the river, Dickey and St. Leonard being the farthest upstream and downstream, respectively (approximate river distances between consecutive locations: 58 km Dickey to Clair, 34 km Clair to Edmundston, and 42 km Edmundston to St. Leonard). The small pulse at Clair on 16 April was preceded by the release of an ice jam located ~39 km upstream. The release occurred at ~1140 h and produced a 0.3 m wave height at the Ft. Kent gauge, which is located 1.3 km downstream of the Clair bridge. The ice from the released jam was arrested again at ~21 km upstream of Clair. The new jam may have advanced closer to Clair after that time and released very late on the 16th or very early on the 17th, producing a 3-m jave at the Fort Kent gauge in the morning of the 17th. The accompanying ice run was starting to thin out at Clair by 0630 h on the 17th and advanced to well beyond St. Leonard, where the largest sediment pulse was recorded on 17 April.

A temporal association exists between sediment pulses and the ice runs that accompany various javes (Figure 10). The pulses are caused by the amplified erosive capacity of the flow during the passage of javes, while as many as three pulses per site have been observed during a breakup event. At a given location, the sediment pulse begins after and ends before the jave does. Depending on the intensity of the ice run that ensues upon ice-jam release, the peak of the pulse may coincide with the peak of the jave (moderate runs) or occur before it (heavy runs). It has been hypothesized that the latter outcome results from near-shore grounding of ice rubble, which shields the river banks from further erosion.

Javes produce much higher concentrations than would be expected from the gradual river response to runoff, ranging from 4.2 to 6.5 times the runoff-generated peaks in the Saint John River and attaining values of ~1000 mg/L or 1 g/L [53]. For the Athabasca River above Fort McMurray, known for highly dynamic breakups, it is estimated that jave-induced SSCs could exceed ~10 g/L [19].

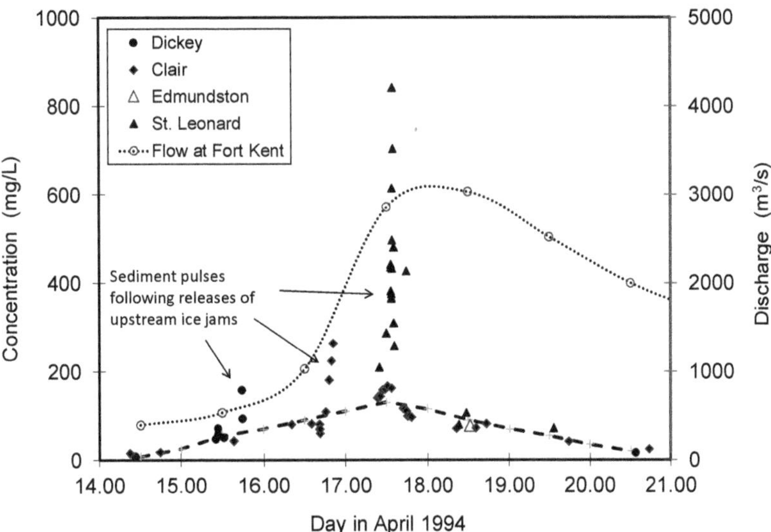

Figure 9. Suspended-sediment concentrations and mean daily flows during the 1994 breakup event in the upper Saint John River. The dashed line approximates the variation of "carrier" (no jave) concentration with time at a gauge located 1.3 km downstream of Clair. From [53], Crown copyright, published by Elsevier.

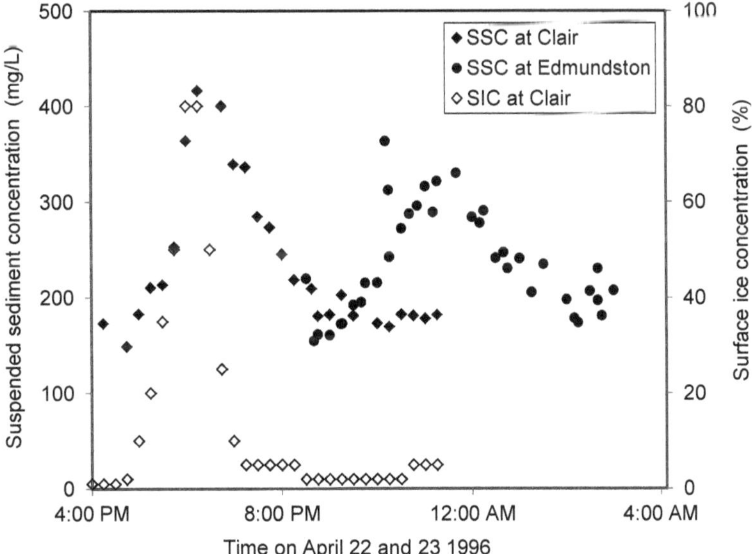

Figure 10. Pulse propagation and dispersion, Saint John River, 22 and 23 April 1996. The Clair bridge is located ~34 river km upstream of the Edmundston bridge. The variation of surface ice concentration (SIC) at Clair is also shown. From [53], Crown copyright, published by Elsevier.

The occurrence of sediment pulses during javes can be explained by increased sediment supply and flow velocity/shear stress upon release of an ice jam. Water levels upstream of ice jams rise, exposing more of the sources of sediment, such as river banks and flood plains, to flowing water. When the jams release, upstream water levels drop quickly, and water re-entering the channel potentially carries with it considerable sediment. Downstream of the former ice-jam location, water levels rise as the jave passes, exposing still-dry bank zones; at the same time, transient but very high flow velocities and shear stresses cause bed and bank erosion. Though javes are typically very brief relative to the duration of the breakup event, they can deliver significant sediment loads, owing to the amplified SSCs and concomitant flows [19,53].

Peak sediment concentrations and net loads delivered by breakup pulses depend on a variety of factors related to jave intensity and sediment availability. The data for the Saint John River at Clair as reported in [53] reveal that pre-pulse and peak pulse SSCs increase with "carrier" discharge, i.e., the runoff-generated discharge in the river, which prevails before or after jave events. However, the amount of scatter associated with peak values suggests that additional factors, such as jam location and channel morphology, also need to be taken into account. We are not aware of any bed load measurements performed during the passage of javes in rivers, but bed shear magnitudes inferred from RLAM applications suggest that bed mobilization and transport can be intense.

Initial quantitative understanding of bed load and suspended bed material transport could be developed by applying existing modelling capability [49,50] and by adaptation of other dynamic river ice models (e.g., River 1d; [65]) to include sediment transport/deposition equations. However, no known river ice model has bank erosion prediction capability.

6. Environmental Implications

6.1. Geomorphic Considerations

The presence of a floating ice cover modifies the magnitude and distribution of flow velocities and thus the boundary and internal shear stresses that affect transport of bed and suspended sediment. Significant scour can occur especially at the toe of an ice jam, where the downward protrusion of ice deflects flow towards the bed and increases velocity through a restricted flow area under the jammed ice. In rivers and streams with mobile bed material, channel geometry and the bed regime can change corresponding to changes in erosive force and sedimentation.

Sediment moving in a river channel can be classified as bedload, suspended load, and mixed load [66]. In suspended-load-dominated systems (suspended sediment 85–100% of total sediment load), floods deposit finer, more cohesive sediments on the flood plain and stream banks, and the channel becomes narrow and deeper and tends to meander. Streambed erosion predominates over channel widening [66]. Bedload-dominated channels (bedload sediment 35–70% of total sediment load) display stream deposition as gravel bars and islands in a relatively wide, steep channel. Erosion occurs mainly from channel widening [66]. Mixed-load dominated channels experience both modes of sedimentation. The characteristics of suspended load-, mixed load- and bedload-dominated channels were expanded upon in [67]. The type of system (bedload, suspended load, or mixed load) can be delineated based on the Shields shear stress and the Reynolds number. The observed properties of the stream channel (channel pattern) can be used to infer sediment transport and depositional processes [68]. A river's response to changes in streamflow and sediment transport can be evaluated using Lane's relationship, where the product of flow discharge and energy slope is proportional to the product of sediment transport and median transport rate [69].

Sediment supply and deposition, including during the winter and ice-breakup period, should be considered when determining the functioning and operational life of a dam and reservoir. Dam construction alters the natural equilibrium between sediment dynamics and morphology along a river, including downstream impacts on river morphology and ecology due to a disruption in sediment continuity [70]. Some degree of sediment inflow and deposition occurs in all reservoirs formed by dams [70,71]. Sedimentation can reduce the longevity, usefulness, and sustainable operations of both storage reservoirs and run-of-the-river projects [72,73]. Generally, coarser sediments deposit in the upper ends of reservoirs, while finer suspended sediments may flow though the outlet works or deposit in the reservoir nearer the dam [70,71]. Factors affecting sediment deposition in reservoirs include the supply of sediment from upstream and the shoreline of the reservoir, sediment characteristics, reservoir dimensions and shape (that affects the flow-velocity distribution), and the outflow or sluicing of sediment at a dam's outlet works. Trap efficiency, the ratio of the quantity of deposited sediment to the total sediment inflow, is affected by the sediment particle fall velocity (dependent upon particle size and shape and upon water viscosity and chemistry) and the rate of flow through the reservoir (subject to reservoir storage and rate of outflow) [71].

Flow regulation can alter channel morphology of ice-affected rivers. Widespread geomorphological changes, including the loss of sinuosity and the widening of the active unvegetated channel, have been observed along the Aishihik River, Yukon, and attributed mainly to the effect of regulation on river-ice processes [74]. In addition to other factors, downstream transport and deposition characteristics of the river may be significantly modified because ice breakup regimes can change considerably under regulation [75]. Such changes should be carefully considered in planning new projects or in assessing environmental impacts of existing projects. Because ice breakup and jamming processes are influenced by antecedent freezeup and winter conditions, in addition to the flow hydrograph, the entire regulated ice regime would have to be studied.

6.2. Water Quality Considerations

Trace element and organic contaminant concentrations will vary in bed sediment depending on the physical and chemical characteristics of the sediment and the concentrations of various contaminants. Sorption capacity of bed materials generally increases with decreasing particle size. Therefore, clay would be expected to have a greater capacity for surface coatings of carbonates, organic matter; and hydrous oxides of iron (Fe), aluminium (Al), and manganese (Mn) than sand [76]. Finer materials are deposited to bed sediments in fluvial systems during the under-ice period, leading to an accompanying increase in sediment TOC content and in the concentrations of a variety of trace elements [76].

Ice jams are important agents affecting sediment metal concentrations in rivers [77]. Following a large ice jam and associated flooding in February 1996 on the Blackfoot and Clark Fork Rivers of western Montana, large amounts of fine-grained sediment were mobilized such that metal concentrations in sediment downstream from a reservoir containing large amounts of contaminated sediment were enriched in metals, while open reaches above the reservoir were diluted [77].

Many water-quality contaminants travel as sorbed constituents on suspended-sediment particles, and thus water quality can be related to the suspended-sediment transport. To examine the concentrations of trace metals, 13 pairs of simultaneous water samples were obtained by dip sampling during the breakup event of 1997 at three bridge sites (Clair, Edmundston and St. Leonard) across the Saint John River, New Brunswick [52]. Environment Canada's National Laboratory for Environmental Testing (NLET) in Burlington, Ontario analysed different samples from each set for dissolved and total concentrations of 17 metals: aluminium (Al), barium (Ba), beryllium (Be), cadmium (Cd), chromium (Cr), cobalt (Co), copper (Cu), iron (Fe), lead (Pb), lithium (Li), manganese (Mn), molybdenum (Mo), nickel (Ni), silver (Ag), strontium (Sr), vanadium (V), and zinc (Zn). For the study stretch of the Saint John River, the highest metal concentrations often occur during the breakup period when suspended-sediment concentrations are highest [52,78,79]. Measured breakup concentrations of aluminium, iron, and copper exceeded measured open-water values potentially by an order of magnitude; however, the highest open-water values of zinc (to 0.12 mg/L) exceeded breakup values [52].

A strong metal association with suspended sediment is evinced as total metal concentration tends to increase with suspended-sediment concentration [52]. For the data obtained on the study stretch of the Saint John River, total metal concentration (unfiltered sample), C_T, could be determined from a linear relationship:

$$C_T = s\,C_s + C_w \tag{7}$$

in which s = particulate metal concentration expressed as mass of metal per unit mass of suspended sediment, C_s = suspended-sediment concentration, and C_w = dissolved metal concentration (filtered sample).

As C_W is usually a small part of C_T, the total metal content is largely governed by the value of the SSC. The particulate metal concentration, s, may be decreasing, increasing, or trendless with respect to discharge [80]. For the SJR data, an increasing value of s with increasing discharge was the general case for copper (Cu) and zinc (Zn), while a decreasing value of s with increasing discharge was found for several other metals (Al, Ba, Cd, Co, Cr, Fe, Li, Mn, Ni, Pb, Sr, and V). Despite the variability of s with discharge, it is still possible to use for practical purposes an appropriate value of s and an average value of C_W to estimate total metal concentration [52].

The surface area of sediment per unit mass and the sorption distribution coefficient are two important factors affecting the variability of the particulate trace metal concentration. The surface area of the sediment particles per unit mass, A_m, reflects the area for sorption and relates to relevant chemical processes [81]. Assuming thin platelets and using an average median size (D_{50}) of about 11 μm for the primary sediment particles, the estimated surface area of the sediment particles per unit mass, A_m, at the metal-sampling sites on the Saint John River is ≈ 7 m^2/g [52]. The sorption distribution coefficient (particulate metal concentration (mass of metal per unit mass of suspended sediment) divided by the dissolved metal concentration) was reported as 5.2, 4.1, 5.0, 4.2, 5.4, 5.6, 2.7, 5.1, and 4.4 for AL, Ba, Cr, Cu, Fe, Mn, Ni, Pb, Sr, V, and Zn, respectively [52]. These values are in approximate agreement with values obtained elsewhere, as compiled in [82].

River ice can also be a depository for various substances, including toxic elements, as evinced by studies of ice in the Amur River, Russia [83]. The deposition of these substances is the result of ice interacting with flowing water and the river bed and banks as well as of the settling of atmospheric particles (including aerosols and suspended particulates) on the ice surface. The freezeup period provides environmental conditions for the accumulation of these substances into the ice cover and for transport, sedimentation, migration, and transformation of biogenic elements, suspended particulates, and dissolved organic matter [83]. Processes occurring during ice-cover growth also may change the migration ability of many heavy metals, thereby creating conditions for buildup of more toxic compounds [83]. During the melting and breakup of ice in springtime, these pollutants can enter the water and are thus unloaded into aquatic ecosystems [83].

The long-term retention and accumulation of metals in riverine aquatic environments can be detrimental to biota and ecosystem processes regardless of hydroclimatic setting and metal source (point or non-point, natural or anthropogenic). Metals on sediment that deposits on the channel bed can become dissolved into the water column over time and possibly bioaccumulate in living organisms. The amount of deposited substance can be effectively quantified by considering load rather than concentration, with a more accurate annual or seasonal estimate of substance loading obtained if sediment processes during breakup are not ignored.

6.3. Ecological Considerations

Scrimgeour et al. [84] reviews the effects of ice-cover breakup on aquatic systems, and Prowse [85,86] describes how river-ice phenomena and processes affect hydrologic, geomorphic, and chemical characteristics of a river and the associated biological consequences. Yet, complete understanding of how winter conditions affect aquatic ecosystems is lacking [87]. Linking changes in ice cover with river ecology is difficult because information about ice in small rivers and about ecology in the large rivers is lacking [88]. In this subsection, consideration is given to the effects of sediment transport and higher SSCs on aquatic species, aquatic habitat, and riverbank species, including vegetation.

River-ice processes change conditions for aquatic life by preventing reaeration (and thus reducing dissolved oxygen levels) in ice-covered rivers; by changing physical and chemical parameters in the under-ice flow, such as light intensity, temperature, stream velocity, and contaminant mixing; by creating a surface barrier that alters the prey–predator relationship, and by changing the amount of sediment transport and deposition. Changes in geomorphology, temperature regimes, and hydraulic connectivity along a river that affect biological and chemical processes can vary from headwaters to river mouth, with these changes likely greater in larger river systems [88]. A direct, physical effect upon biota can be more likely in small to medium streams of shallow depth than larger deeper rivers due to the greater possibility in small to medium streams of ice damming (freezing to the channel bed), anchor-ice formation, and ice scouring of the channel bed [88]; the latter two processes potentially add sediment to the river flow.

Aquatic biota respond to both the concentration of suspended sediments and duration of exposure, much as they do for other environmental contaminants; regression analysis indicates that SSC alone is a relatively poor indicator of effects on aquatic life [89]. The severity of the effects of suspended sediment also depends upon taxonomic group, life stage, and life history of the affected biota; the particle size, angularity, and toxicity of the suspended sediment; and duration of exposure to SSCs above normal background levels. In most cases, elevated suspended sediments cause sublethal effects, as some species can tolerate very high levels of suspended solids within a short period of time. However, harmful effects in ultrasensitive species and in species during certain life stages may occur once a threshold concentration of suspended sediment is abruptly exceeded [90]. Elevated suspended-sediment concentrations can lead to stress that increases vulnerability to other environmental stressors, such as water pollutants.

Considerable evidence exists to indicate that changes in the abundance and composition of invertebrate communities are associated with increases in suspended solids and turbidity [91,92]. Sediment can also directly affect invertebrates by clogging the filter mechanisms of bottom feeders or harming by abrasion the gills impairing respiration [89]. Excess deposition of sediment in coarse substrates will cover and inundate the gravels, thereby eliminating the habitat where many invertebrates live [91–94]. Ice regimes in temperate rivers can also affect organic-matter dynamics and feeding ecology of aquatic insects [95].

Elevated levels of suspended sediments can impact fish by physically damaging tissues and organs. Fine sediment can easily damage fish gills, with angular sediment particles usually more damaging to gills than larger or rounded particles. The scouring and abrasive action of suspended particles can damage gill tissues or clog gills [96]. If fish gills are irritated over sufficient time, mucus that forms to protect the gill can impede respiration [90,96–98]. Physical damage and clogging of the gills can increase fish susceptibility to infection and disease, reduce growth rates, and even result in fish mortality if high SSCs persist over a prolonged period [90,96]. Although it depends upon the species and life stage, moderate and severe gill damage may occur when SSCs exceed 100 mg/L and 500 mg/L, respectively [96]. Sediment mineralogy and the presence of innate or adsorbed toxicants may affect fish at the tissue and cellular level [90]. Kemp et al. [92] review the impacts of fine sediment on fish physiology and performance, including stream and sediment characteristics that affect the effect of fine sediment on fish (Table 1 of [92]) and the temporal responses of fish to fine sediment (Table 2 of [92]). Table 1 of [99] identifies the mortality effects of suspended-sediment levels by fish species. Affandi and Ishak [98] reviews the effects of turbidity, suspended sediment, and metal pollution on fish populations and riverine ecosystem, noting that the health of fish and their abundance also represent the health of their corresponding water bodies.

Elevated levels of suspended sediments can also impact fish by increasing physiological stress. The physiological stress caused by exposure to elevated SSCs over time can make fish more susceptible to disease [96]. The physiological effects of suspended sediments on fish are influenced by water temperatures, as dissolved oxygen is higher, and fish exhibit reduced activity (and metabolic rate) at cooler temperatures [91].

Suspended sediments also indirectly affect fish by changing water clarity (increased turbidity), which can alter fish movement or migration patterns, feeding success, and habitat quality [96]. Suspended solids have significant effects on community dynamics when they interfere with light transmission because of turbidity [93]. These factors also create stress levels that can trigger, over time, damaging physiological effects. Several researchers have investigated the effects of SSCs on fish avoidance of areas of high SSC by moving to areas of lower SSC. Knack and Shen [100] reviews existing studies on fish behaviour during the winter in ice-affected climates. Fish, particularly salmon, are sight feeders [91]. Turbidity decreases the ability of fish to find food and this, over time, may affect their health and rate of growth. Turbid water results in a stress response in salmonids.

Aggregates of fine sediment containing organic material is a primary source of food and energy for river ecosystems. The ability of taxa to adapt to variability in suspended loads and turbidity under natural conditions characterize natural ecosystems, creating opportunities for some species and affecting the structure of biological communities [92]. Short-term physiological and behavioural effects of fish affect the survival of a local fish population to high suspended-sediment loads; however, such exposure events may also have intergenerational ramifications that affect fish condition, survival, populations, and communities [99]. Recovery of a fish population from a sediment pulse depends on the magnitude and period of the sediment perturbation and sediment composition [92].

International guidelines for suspended sediment vary. The Canadian Water Quality Guidelines for the Protection of Aquatic Life provide target limits for total particulate matter in freshwater, estuarine, and marine environments with respect to suspended sediments, turbidity, deposited bedload sediments, and streambed substrate [101]. In clear stream systems, small, induced exceedances in suspended-sediment concentration above a 25 mg/L change from background levels for short-term exposure (e.g., 24 h) can cause behavioural and low sublethal effects on fish, all of which are reversible (CCME 2002). During high flows, an increase of 25 mg/L when sediment levels are between 25 mg/and 250 mg/L or by 10% or more when sediment levels exceed 250 mg/L is considered detrimental to aquatic life. Information in technical literature as cited in this paper shows that such increases can occur during breakup. For longer term exposure (e.g., 30 d or more), an increase in average suspended-sediment concentrations by more than 5 mg/L over background levels is identified as of concern [101]. In Europe, suspended-sediment concentrations should not exceed 25 mg/L for salmonid and cyprinid habitats [96].

Exposure to certain chemical substances represents a potentially significant hazard to the health of the organisms. Sediment quality guidelines contain information regarding the relationships between the chemical concentrations associated with sediment and adverse biological effects resulting from exposure to these chemicals [101].

Habitat requirements shift with ontogenetic development [92]. Fish eggs are an especially vulnerable life stage due to their immobility and the requirements for specific hydraulic and substrate conditions [102]. Deposition and the infilling of gravel bed material by finer sediment can reduce survival of fish eggs by reducing clean water flow and oxygen from reaching the eggs and by preventing newly hatched fish from migrating through the gravel and emerging as fry [91]. Generally, stable discharges and low sediment discharge are more favourable to the survival of salmonids in the early stages of life than sudden changes in flow and sediment discharge. Small amounts of fine sediment in salmonid redds can diminish survivability and emergence of fish from eggs. The intrusion of fine sediments into a gravel streambed depends upon the initial concentration of suspended sediment; the roughness and particle size distribution of the bed material; and flow characteristics, such as discharge, water depth, velocity, and shear stress [103]. Kemp et al. [92] discussed the spawning traits and incubation periods of several benthic spawning freshwater fish, detailing the spawning habitat, egg characteristics, incubation period, and potential impacts of fine sediments (Table 3 of [92]). As is the case for eggs, larval and juvenile stages are more susceptible to the effect of fine sediment than adults, which have a greater propensity to move and avoid adverse conditions [104].

Hydro-morphological conditions in rivers can affect fish habitat, which can have important effects on fish mortality during the winter months. Knack and Shen [100] presents proposed fish-habitat preferences for salmonids in winter. Habitat suitability curves were developed for use with a two-dimensional, hydro-thermal-ice-sediment river dynamics model, RICE2D. They simulated river conditions with the RICE2D model coupled with an IFIM and PHABSIM-type fish-habitat-suitability model, which includes thermal-ice as well as sediment and bed-change conditions in rivers to allow a detailed habitat suitability evaluation. The impact of the dam removal on the St. Regis River, Hogansburg, NY, USA, on fish habitat was evaluated, showing an improvement in fish habitat for both open-water and ice-affected conditions in the restored reach. Model results show that there would be no significant overall loss of wintertime fish habitat from the removal of the dam despite increased frazil ice production [100]. The model is a useful assessment tool for exploring fish habitat during the winter ice periods and for evaluating fish-habitat restoration for all seasons.

In addition to aquatic fauna, the geomorphologic processes associated with ice breakup (see Section 3.1) also affect vegetation. The amount of ice-related stress or disturbance that a plant species can tolerate and its capability to recover after an ice-related disturbance depends upon its physiology and the nature of the disturbance [105]. Tree scars, a phenomenon that occurs when ice shoves against trees along the riverbank, and trim lines, the lower limit of a vegetative community longitudinally along a river, can provide information on the elevation of past ice-jam events (e.g., [105]). In some cases, ice scour and the transport of sediment and organic matter can contribute to the dissemination and available habitat of plant species. For example, the population of Furbish's lousewort (Pedicularis furbishiae), an endangered herbaceous perennial plant that occurs on the intermittently flooded 140-km long stretch of Saint John River in northern Maine and north western New Brunswick, is dependent on the presence of appropriate host plants, moist soils, and pollinators and on habitat created by periodic ice scour, providing the occurrence of ice-related scour is not too frequent or rare [106].

In addition, ice-jam-related flooding and replenishment with water, sediment, and nutrients can be critical to the ecosystem vitality of northern river-delta environments, such as the Mackenzie River Delta and the Peace-Athabasca Delta in northern Canada. This is particularly concerning if perched ponds and lakes are too high to be recharged by the open-water flow system [107,108].

7. Concluding Remarks

Compared to midwinter periods under a stable ice cover, forces acting to dislodge material from a river bank during the pre-breakup and breakup periods not only increase, but rising water levels may expose a greater portion of the bank to these forces. This results in increased sediment transport, which can have potential effects on channel morphology, reservoir operations, water quality, aquatic life and habitat, and riparian ecology.

Under a changing and progressively warmer climate, an increase in the number of midwinter breakups and the magnitude of breakup flows may result in a considerable increase in future sediment concentrations and loads during the pre-breakup and breakup periods. Projections of regional climate change need to be inputted in hydrologic and hydrodynamic models to determine the physical changes in surface runoff and streamflow that may alter sediment supply, transport, and deposition in rivers with seasonal ice covers.

Many northern rivers exhibit low background suspended-sediment concentrations prior to breakup, but suspended-sediment concentrations increase by at least an order of magnitude prior to and during breakup. Suspended-sediment concentrations during javes can be even greater and far exceed those of open-water events. Therefore, to estimate the total annual load in a river, it is necessary to consider the amount of suspended sediment delivered during the breakup period. An increase in suspended-sediment concentrations during breakup is of interest due to their effect on aquatic ecosystems in terms of both species' life cycle and habitat and on water quality due to the sorption of particulate contaminants onto sediment particles.

Scientific studies have shown that suspended-sediment loads during the breakup period may be a major portion of the annual sediment load on some rivers. Yet, existing sediment monitoring programs normally do not capture the increased suspended-sediment concentrations and loads that occur briefly during the ice breakup event. A more accurate estimate of sediment loads on a river may be obtainable from more frequent measurements of suspended sediment during the breakup period, but large-scale collection of sediment data during breakup is likely to be difficult and costly.

As revealed in scientific literature, high sediment concentrations, especially when associated with metal contaminants, can have detrimental effects on water quality and ecosystem robustness. Total metal concentration tends to increase with suspended-sediment concentration. Along a stretch of the Saint John River, high metal concentrations were found generally to occur during the breakup period when suspended-sediment concentrations are high.

Further research is needed to relate sediment concentrations and loads, including those generated by sediment pulses, to breakup conditions and underlying hydro-climatic and geomorphological variables and hence develop predictive mathematical models. Affordable instrumentation and data-interpretation software could also be developed and applied to obtain adequate field datasets for model calibration.

Author Contributions: S.B. and B.C.B., conceptualization, writing—review and editing. All authors have read and agreed to the published version of the manuscript.

Funding: This research received no external funding.

Institutional Review Board Statement: Not applicable.

Informed Consent Statement: All individuals included in this section have consented to the acknowledgement.

Data Availability Statement: Not applicable.

Acknowledgments: Support by Environment and Climate Change Canada to produce this article, and helpful discussion with Ian Knack are acknowledged with thanks.

Conflicts of Interest: The authors declare no conflict of interest.

References

1. Burrell, B.C.; Beltaos, S. Effects and Implications of River Ice Breakup on Suspended-Sediment Concentrations: A synthesis. In Proceedings of the CGU HS Committee on River Ice Processes and the Environment 20th Workshop on the Hydraulics of Ice-Covered Rivers, Ottawa, ON, Canada, 14–16 May 2019. Available online: http://www.cripe.ca/docs/proceedings/20/Burrell-Beltaos-2019.pdf (accessed on 10 September 2021).
2. Beltaos, S. (Ed.) *River Ice Breakup*; Water Resources Publications, LLC: Highlands Ranch, CO, USA, 2008; 462p.
3. Beltaos, S. Hydrodynamic characteristics and effects of river waves caused by ice jam releases. *Cold Reg. Sci. Technol.* **2013**, *85*, 42–55. [CrossRef]
4. Turcotte, B.; Burrell, B.C.; Beltaos, S. The Impact of Climate Change on Breakup Ice Jams in Canada: State of knowledge and research approaches. In Proceedings of the 20th Workshop on the Hydraulics of Ice Covered Rivers, Ottawa, ON, Canada, 14–16 May 2019. Available online: http://www.cripe.ca/docs/proceedings/20/Turcotte-et-al-2019.pdf (accessed on 15 September 2021).
5. Beltaos, S.; Prowse, T. River-ice hydrology in a shrinking cryosphere. *Hydrol. Process.* **2009**, *23*, 122–144. [CrossRef]
6. Kämäri, M.; Alho, P.; Veijalainen, N.; Aaltonen, J.; Huokuna, M.; Lotsari, E. River ice cover influence on sediment transportation at present and under projected hydroclimatic conditions. *Hydrol. Process.* **2015**, *29*, 4738–4755. [CrossRef]
7. Chassiot, L.; Lajeunesse, P.; Bernier, J.-F. Riverbank erosion in cold environments: Review and outlook. *Earth-Sci. Rev.* **2020**, *207*, 103231. [CrossRef]
8. Gatto, L.W. Riverbank Conditions and Erosion during Winter. In Proceedings of the Workshop on Environmental Aspects of River Ice, Saskatoon, SK, Canada, 18–20 August 1993; Prowse, T.D., Ed.; NHRI Symposium Series No. 12; National Hydrology Research Institute: Saskatoon, SK, Canada, 1993; pp. 43–56.
9. Vandermause, R.A. The Role of Dynamic Ice-Breakup on Bank Erosion and Lateral Migration of the Middle Susitna River, Alaska. Master's Thesis, Department of Civil and Environmental Engineering, Colorado State University, Fort Collins, CO, USA, 2018; 334p. Available online: https://mountainscholar.org/handle/10217/191335?show=full (accessed on 10 September 2021).

10. Vandermause, R.; Harvey, M.; Zevenbergen, L.; Ettema, R. River-ice effects on bank erosion along the middle segment of the Susitna river, Alaska. *Cold Reg. Sci. Technol.* **2021**, *185*, 103239. [CrossRef]
11. Vandermause, R.; Ettema, R. Aerial Monitoring of Vegetation to Identify Ice-related Bank Erosion Along the Middle Reach of the Susitna River, Alaska. In Proceedings of the CGU HS Committee on River Ice Processes and the Environment 19th Workshop on the Hydraulics of Ice Covered Rivers, Whitehorse, YK, Canada, 9–12, July 2017.
12. Ettema, R.; Kempema, E.W. Chapter 9: Ice effects on sediment transport. In *River Ice Formation*; Beltaos, S., Ed.; Committee on River Ice Processes and the Environment, Canadian Geophysical Union Hydrology Section: Edmonton, AB, Canada, 2013; pp. 297–338.
13. Turcotte, B.; Morse, B.; Bergeron, N.E.; Roy, A.G. Sediment transport in ice-affected rivers. *J. Hydrol.* **2011**, *409*, 561–577. [CrossRef]
14. Ettema, R. Review of Alluvial-Channel responses to River Ice. *J. Cold Reg. Eng.* **2002**, *16*, 191–217. [CrossRef]
15. Gatto, L.W. *Soil Freeze-Thaw Effects on Bank Erodibility and Stability*; CRREL Special Report 95-24; Cold Regions Research and Engineering Laboratory: Hanover, NH, USA, 1995; 24p.
16. Chow, V.T. *Open Channel Hydraulics*; McGraw-Hill Book Company: New York, NY, USA, 1959.
17. Yalin, M.S. *Mechanics of Sediment Transport*; Pergamon Press: Oxford, UK, 1977; 360p, ISBN 978-0080211626.
18. Beltaos, S.; Carter, T.; Rowsell, R.; DePalma, S.G.S. Erosion potential of dynamic ice breakup in Lower Athabasca River. Part I: Field measurements and initial quantification. *Cold Reg. Sci. Technol.* **2018**, *149*, 16–28. [CrossRef]
19. Beltaos, S. Erosion potential of dynamic ice breakup in Lower Athabasca River. Part II: Field data analysis and interpretation. *Cold Reg. Sci. Technol.* **2018**, *148*, 77–87. [CrossRef]
20. Annandale, G.W. Erodibility. *J. Hydraul. Res.* **1995**, *33*, 471–494. [CrossRef]
21. Turcotte, B.; Morse, B. The Winter Environmental Continuum of Two Watersheds. *Water* **2017**, *9*, 337. [CrossRef]
22. Hicklin, E.J. Chapter 4: Sediment transport. In *Rivers. Course Notes for an Undergraduate Course in Fluvial Geomorphology*; Department of Geography, Simon Fraser University: Burnaby, BC, Canada, 2009. Available online: http://www.sfu.ca/~{}hickin/RIVERS/Rivers4%28Sediment%20transport%29.pdf (accessed on 16 March 2021).
23. Knighton, D. *Fluvial Forms and Processes: A New Perspective*; Arnold: London, UK, 1998; 383p.
24. Robert, A. *River Processes: An Introduction to Fluvial Dynamics*; Arnold: London, UK, 2003; ISBN 978-0340763391.
25. Van Rijn, L.C. Sediment Transport, Part II: Suspended Load Transport. *J. Hydraul. Eng.* **1984**, *110*, 1613–1641. [CrossRef]
26. Tsujimoto, T. Diffusion coefficient of suspended sediment and kinematic eddy viscosity of flow containing suspended load. In *River Flow 2010*; Dittrich, A., Koll, K., Aberle, J., Geisenhainer, P., Eds.; Bundesanstalt für Wasserbau S.: Karlsruhe, Germany, 2010; pp. 801–806.
27. Syvitski, J.; Cohen, S.; Miara, A.; Best, J. River temperature and the thermal-dynamic transport of sediment. *Glob. Planet. Chang.* **2019**, *178*, 168–183. [CrossRef]
28. Fondriest Environmental, Inc. Turbidity, Total Suspended Solids and Water Clarity. Fundamentals of Environmental Measurements. 13 June 2014. Available online: https://www.fondriest.com/environmental-measurements/parameters/water-quality/turbidity-total-suspended-solids-water-clarity/ (accessed on 10 September 2021).
29. Kempema, E.W.; Ettema, R.; Bunte, K. Anchor-ice rafting, an important mechanism for bed-sediment transport in cold-regions rivers. In *River Flow 2020*; Taylor and Francis: Oxfordshire, UK, August 2020. [CrossRef]
30. Kalke, H.; McFarlane, H.; Schneck, C.; Loewen, M. The transport of sediments by released anchor ice. *Cold Reg. Sci. Technol.* **2017**, *143*, 70–80. [CrossRef]
31. Maclean, B.; Bampfylde, C.; Lepine, M.; Tssessaze, L. Towards a Rights-Based Ice Monitoring Trigger. In Proceedings of the 21st Workshop on the Hydraulics of Ice Covered Rivers, Saskatoon, SK, Canada, 29 August–1 September 2021.
32. Lawson, D.E.; Chacho, E.F., Jr.; Brockett, B.E.; Wuebben, J.L.; Collins, C.M.; Arcone, S.A.; Delaney, A.J. *Morphology, Hydraulics and Sediment Transport of an Ice-Covered River: Field Techniques and Initial Data*; CRREL Report 86-11; U.S. Army Cold Regions Research and Engineering Laboratory: Hanover, NH, USA, 1986; 49p.
33. Prowse, T.D. Suspended sediment concentration during breakup. *Can. J. Civ. Eng.* **1993**, *20*, 872–875. [CrossRef]
34. Beltaos, S.; Burrell, B.C. Transport of suspended sediment during the breakup of the ice cover, Saint John River, Canada. *Cold Reg. Sci. Technol.* **2016**, *129*, 1–13. [CrossRef]
35. Ferrick, M.G.; Weyrick, P.B. On the Sediment Carrying Capacity of Rivers during Ice Breakup. In Proceedings of the Workshop on Environmental Aspects of River Ice, Saskatoon, SK, Canada, 18–20 August 1993; Prowse, T.D., Ed.; NHRI Symposium Series No. 12; National Hydrology Research Institute: Saskatoon, SK, Canada, 1993; pp. 161–175.
36. Beltaos, S. Chapter 7. Freezeup jamming and formation of ice cover. In *River Ice Formation*; Beltaos, S., Ed.; CGU-HS Committee on River Ice Processes and the Environment (CRIPE): Edmonton, AB, Canada, 2013; pp. 181–255.
37. Gerard, R. Preliminary observations of spring ice jams in Alberta. In Proceedings of the International Association for Hydraulic Research Ice Symposium, Hanover, NH, USA, 18–21 August 1975; pp. 261–277.
38. Andres, D.D.; Rickert, H.A. *Observations of the 1985 Breakup in the Athabasca River Basin Upstream of Fort McMurray, Alberta*; Report No. SWE 85-10; Alberta Research Council: Edmonton, AB, Canada, 1985; 74p.
39. Jasek, M.; Beltaos, S. Chapter 8. Ice-jam release: Javes, ice runs and breaking fronts. In *River Ice Breakup*; Water Resources Publications: Highlands Ranch, CO, USA, 2008.
40. She, Y.; Hicks, F. Modelling ice jam release waves with consideration for ice effects. *Cold Reg. Sci. Technol.* **2006**, *45*, 137–147. [CrossRef]

41. Kowalczyk Hutchison, T.; Hicks, F.E. Observations of ice jam release waves on the Athabasca River near Fort McMurray, Alberta. *Can. J. Civ. Eng.* **2007**, *34*, 473–484. [CrossRef]
42. She, Y.; Andrishak, R.; Hicks, F.; Morse, B.; Stander, E.; Krath, C.; Keller, D.; Abarca, N.; Nolin, S.; Nzokou Tanekou, F.; et al. Athabasca River ice jam formation and release events in 2006 and 2007. *Cold Reg. Sci. Technol.* **2009**, *55*, 249–261. [CrossRef]
43. Nafziger, J.; She, Y.; Hicks, F. Celerities of waves and ice runs from ice jam releases. *Cold Reg. Sci. Technol.* **2016**, *123*, 71–80. [CrossRef]
44. Beltaos, S.; Burrell, B.C. Field measurements of ice-jam-release surges. *Can. J. Civ. Eng.* **2005**, *32*, 699–711. [CrossRef]
45. Henderson, F.M. *Open Channel Flow*; The Macmillan Company: New York, NY, USA, 1996; 544p, ISBN 978-0023535109.
46. Beltaos, S.; Burrell, B.C. Determining ice-jam surge characteristics from measured wave forms. *Can. J. Civ. Eng.* **2005**, *32*, 687–698.
47. Beltaos, S.; Rowsell, R.; Tang, P. Remote data collection on ice breakup dynamics: Saint John River case study. *Cold Reg. Sci. Technol.* **2011**, *67*, 135–145. [CrossRef]
48. Knack, I.M.; Shen, H.T. Numerical Modeling of Ice Transport in Channels with River Restoration Structures. *Can. J. Civ. Eng.* **2017**, *44*, 813–819. [CrossRef]
49. Knack, I.M.; Shen, H.T. A Numerical Model for Sediment Transport and Bed Change with River Ice. *J. Hydraul. Res.* **2018**, *56*, 844–856. [CrossRef]
50. Manolidis, M.; Katopodes, N. Bed Scouring During the Release of an Ice Jam. *J. Mar. Sci. Eng.* **2014**, *2*, 370–385. [CrossRef]
51. Knack, I.; Clarkson University, Potsdam, NY, USA. Personal communication, March 2021.
52. Beltaos, S.; Burrell, B.C. Characteristics of suspended sediment and metal transport during ice breakup, Saint John River, Canada. *Cold Reg. Sci. Technol.* **2016**, *123*, 164–176. [CrossRef]
53. Beltaos, S. Extreme sediment pulses during ice breakup, Saint John River, Canada. *Cold Reg. Sci. Technol.* **2016**, *128*, 38–46. [CrossRef]
54. Lai, Y.G. Modeling Stream Bank Erosion: Practical Stream Results and Future Needs. *Water* **2017**, *9*, 950. [CrossRef]
55. Milburn, D.; Prowse, T.D. Observations on Some Physical-Chemical Characteristics of River-Ice Breakup. *J. Cold Reg. Eng.* **2000**, *14*, 214–223. [CrossRef]
56. Milburn, D.; Prowse, T.D. The effect of river-ice breakup on suspended sediment and select trace-element fluxes. *Nord. Hydrol.* **1996**, *27*, 69–84. [CrossRef]
57. Sui, J.; Wang, D.; Karney, B.W. Suspended sediment concentration and deformation of riverbed in a frazil jammed reach. *Can. J. Civ. Eng.* **2000**, *27*, 1120–1129. [CrossRef]
58. Toniolo, H.; Vas, D.; Prokein, P.; Kenmitz, R.; Lamb, E.; Brailey, D. Hydraulic characteristics and suspended sediment loads during spring breakup in several streams located on the National Petroleum Reserve in Alaska, USA. *Nat. Resour.* **2013**, *4*, 220–228. [CrossRef]
59. Moore, S.A.; Zare, S.G.A.; Rennie, C.D.; Seidou, O.; Ahmari, H. Monitoring suspended sediment under ice using acoustic instruments: Presentation of measurements made under break-up. In Proceedings of the 17th Workshop on River Ice, Committee on River Processes and the Environment, Edmonton, AB, Canada, 21–24 July 2013.
60. Shakibaeinia, A.; Dibike, Y.B.; Kashyap, S.; Prowse, T.D.; Droppo, I.G. A numerical framework for modelling sediment and chemical constituents transport in the Lower Athabasca River. *J. Soils Sedim.* **2017**, *17*, 1140–1159. [CrossRef]
61. Skolasińska, K.; Nowak, B. What factors affect the suspended sediment concentrations in rivers? A study of the upper Warta River (Central Poland). *River Res. Appl.* **2018**, *34*, 112–123. [CrossRef]
62. Kowalczyk, T.; Hicks, F. Observations of dynamic ice jam release on the Athabasca River at Fort McMurray, AB. In Proceedings of the 12th Workshop on the Hydraulics of Ice Covered Rivers, Edmonton, AB, Canada, 19–20 June 2003; CGU HS Committee on River Ice Processes and the Environment: Edmonton, AB, Canada, 2003; pp. 369–392. Available online: http://cripe.ca/docs/proceedings/12/Kowalczyk-Hicks-2003.pdf (accessed on 10 September 2021).
63. Malcovish, C.D.; Andres, D.D.; Mostert, P. *Observations of breakup on the Athabasca River near Fort McMurray, 1986 and 1987*; Report SWE-88/12; Alberta Research Council: Edmonton, AB, Canada, 1988.
64. Beltaos, S.; Kääb, A. Estimating river discharge during ice breakup from near-simultaneous satellite imagery. *Cold Reg. Sci. Technol.* **2014**, *98*, 35–46. [CrossRef]
65. Blackburn, J.; She, Y. A comprehensive public-domain river ice process model and its application to a complex natural river. *Cold Reg. Sci. Technol.* **2019**, *163*, 44–58. [CrossRef]
66. Schumm, S.A. *A Tentative Classification of Alluvial River Channels*; Geological Survey Circular 477; US Geological Survey, United States Department of the Interior: Washington, DC, USA, 1963.
67. Church, M. Bed material transport and the morphology of alluvial river channels. *Annu. Rev. Earth Planet. Sci.* **2006**, *34*, 325–354. [CrossRef]
68. Burge, L.; Guthrie, R.; Chaput-Desrochers, L. *Hydrological Factors Affecting Spatial and Temporal Fate of Sediment in Association with Stream Crossings of the Mackenzie Gas Pipeline*; Canadian Science Advisory Secretariat Research Document 2014/029; Fisheries and Oceans Canada: Ottawa, ON, USA, 2014; 35p.
69. Lane, E.W. The importance of fluvial morphology in hydraulic engineering. *Proc. Am. Soc. Civil Eng.* **1955**, *81*, 1–17.
70. Schleiss, A.J.; Franca, M.J.; Juez, C.; De Cesare, G. Reservoir sedimentation. *J. Hydraul. Res.* **2016**, *54*, 595–614. [CrossRef]
71. Strand, R.I.; Pemberton, E.L. *Reservoir Sedimentation Technical Guideline for Bureau of Reclamation*; Engineering and Research Center, Bureau of Reclamation: Denver, CO, USA, 1982; 48p.

72. Palmieri, A.; Shah, F.; Annandale, G.W.; Dinar, A. *Reservoir Conservation: The RESCON Approach*; World Bank: Washington, DC, 2003. Available online: http://documents1.worldbank.org/curated/pt/819541468138875126/pdf/349540v10Reservoir0 conservation0RESCON.pdf (accessed on 25 February 2019).
73. Annandale, G.W.; Morris, G.L.; Karki, P. *Extending the Life of Reservoirs: Sustainable Sediment Management for Dams and Run-of-River Hydropower. Directions in Development*; World Bank: Washington, DC, USA, 2016. [CrossRef]
74. McParland, D.; McKillop, R.; Pearson, F. Fluvial geomorphological responses to river ice dynamics downstream of a hydroelectric facility, southwest Yukon. In Proceedings of the Committee on River Processes and the Environment 21st Workshop on Hydraulics of River Ice, Saskatoon, SK, Canada, 29 August–2 September 2021.
75. Huokuna, M.; Morris, M.; Beltaos, S.; Burrell, B.C. Ice in reservoirs and regulated rivers. *Int. J. River Basin Manag.* **2020**. [CrossRef]
76. Doig, L.E.; Carr, M.K.; Meissner, A.G.N.; Jardine, T.D.; Jones, P.D.; Bharadwaj, L.; Lindenschmidt, K.-E. Open-water and under-ice seasonal variations in trace element content and physicochemical associations in fluvial bed sediment. *Environ. Toxicol. Chem.* **2017**, *36*, 2916–2924. [CrossRef]
77. Moore, J.N.; Landrigan, E.M. Mobilization of metal-contaminated sediment by ice-jam floods. *Environ. Geol.* **1999**, *37*, 96–101. [CrossRef]
78. Beltaos, S.; Burrell, B.C. Suspended sediment concentrations in the Saint John River during ice breakup. In Proceedings of the 2000 Annual Conference of the Canadian Society for Civil Engineering, London, ON, Canada, 7–10 June 2000; pp. 235–242.
79. Beltaos, S.; Burrell, B.C.; Ismail, S. Ice and sedimentation processes in the Saint John River, Canada. In Proceedings of the IAHR International Symposium on Ice, Trondheim, Norway, 11–21 August 1994; Volume 1, pp. 11–21.
80. Miller, J.R.; Orbock Miller, S.M. *Contaminated Rivers*; Springer: Dordrecht, The Netherlands, 2007; 418p.
81. Horowitz, A.J. *A Primer on Sediment-Trace Element Chemistry*, 2nd ed.; Lewis Publishers, Inc.: Chelsea, MI, USA, 1991.
82. Allison, J.D.; Allison, T.L. *Partition Coefficients for Metals in Surface Water, Soil, and Waste*; EPA/600/R-05/074; U.S. Environmental Protection Agency: Washington, DC, USA, 2005; 93p.
83. Golubeva, E.M.; Kondratyeva, L.M.; Shtareva, A.V.; Krutikova, V.O. Features of toxic element distribution in the Amur River Ice. *Earth's Cryosphere* **2020**, *XXIV*, 3–13. [CrossRef]
84. Scrimgeour, G.J.; Prowse, T.D.; Culp, J.M.; Chambers, P.A. Ecological effects of river ice break-up: A review and perspective. *Freshw. Biol.* **1994**, *32*, 261–275. [CrossRef]
85. Prowse, T.D. River-ice ecology. I. Hydrologic, geomorphic, and water-quality aspects. *J. Cold Reg. Eng.* **2001**, *15*, 1–16. [CrossRef]
86. Prowse, T.D. River-ice ecology. II. Biological aspects. *J. Cold Reg. Eng.* **2001**, *15*, 17–33. [CrossRef]
87. Lindenschmidt, K.-E.; Baulch, H.M.; Cavaliere, E. River and Lake Ice Processes—Impacts of Freshwater Ice on Aquatic Ecosystems in a Changing Globe. *Water* **2018**, *10*, 1586. [CrossRef]
88. Thellman, A.; Jankowski, K.J.; Hayden, B.; Yang, X.; Dolan, W.; Smits, A.P.; O'Sullivan, A.M. The ecology of river ice. *J. Geophys. Res. Biogeosci.* **2021**, *126*, e2021JG006275. [CrossRef]
89. Newcombe, C.P.; Macdonald, D.D. Effects of Suspended Sediments on Aquatic Ecosystems. *N. Am. J. Fish. Manag.* **1991**, *11*, 72–82. [CrossRef]
90. Newcombe, C.P.; Jensen, J.O.T. Channel Suspended Sediment and Fisheries: A Synthesis for Quantitative Assessment of Risk and Impact. *N. Am. J. Fish. Manag.* **1996**, *16*, 693–727. [CrossRef]
91. Kerr, S.J. *Silt, Turbidity and Suspended Sediments in the Aquatic Environment: An Annotated Bibliography and Literature Review*; Southern Region Science & Technology Transfer Unit Technical Report TR-008; Ontario Ministry of Natural Resources: Toronto, ON, Canada, 1995; 277p.
92. Kemp, P.; Sear, D.; Collins, A.; Naden, P.; Jones, I. The impacts of fine sediment on riverine fish. *Hydrol. Process.* **2011**, *25*, 1800–1821. [CrossRef]
93. Sorensen, D.L.; McCarthy, M.M.; Middlebrooks, E.J.; Porcella, D.B. *Suspended and Dissolved Solids Effects on Freshwater Biota: A Review*; Corvallis Environmental Research Laboratory, Office of Research and Development, U.S. Environmental Protection Agency: Corvallis, OR, USA, 1977; 73p.
94. Gammon, J.R. *The Effect of Inorganic Sediment on Stream Biota*; Water Pollution Control Research Series 18050 DWC 12/70; Environmental Protection Agency: Washington, DC, USA, 1970; 141p.
95. Blackadar, R.J.; Baxter, C.V.; Davis, J.M.; Harris, H.E. Effects of river ice break-up on organic-matter dynamics and feeding ecology of aquatic insects. *River Res Appl.* **2020**, *36*, 480–491. [CrossRef]
96. Cavanagh, J.-E.; Hogsden, K.L.; Harding, J.S. *Effects of Suspended Sediment on Freshwater*; Landcare Research: Lincoln, New Zealand, 2014; 24p.
97. Berg, L. The effect of exposure to short-term pulses of suspended sediment on the behaviour of juvenile salmonids. In Proceedings of the Carnation Creek Workshop: A Ten-Year Review, Nanaimo, BC, Canada, 24–26 February 1982; Hartman, G.F., Ed.; Department of Fisheries and Oceans, Pacific Biological Station: Nanaimo, BC, Canada, 1982; pp. 177–196.
98. Affandi, F.A.; Ishak, M.Y. Impacts of suspended sediment and metal pollution from mining activities on riverine fish population—A review. *Environ. Sci. Pollut. Res.* **2019**, *26*, 16939–16951. [CrossRef]
99. Kjelland, M.E.; Woodley, C.M.; Swannack, T.M.; Smith, D.L. A review of the potential effects of suspended sediment on fishes: Potential dredging-related physiological, behavioral, and transgenerational implications. *Environ. Syst. Decis.* **2015**, *35*, 334–350. [CrossRef]
100. Knack, I.M.; Shen, H.T. Modeling fish habitat condition in ice affected rivers. *Cold Reg. Sci. Technol.* **2020**, *176*, 103086. [CrossRef]

101. CCME (Canadian Council of Ministers of the Environment). Canadian water quality guidelines for the protection of aquatic life: Total particulate matter. In *Canadian Environmental Quality Guidelines, 1999*; Canadian Council of Ministers of the Environment: Winnipeg, MB, Canada, 2002.
102. Cunjak, R.A.; Prowse, T.D.; Parrish, D.L. Atlantic salmon (Salmo salar) in winter: "The season of parr discontent?". *Can. J. Fish. Aquat. Sci.* **1998**, *55* (Suppl. S1), 161–180. [CrossRef]
103. Glasbergen, K.A. The Effect of Coarse Gravel on Cohesive Sediment Entrapment. Master's Thesis, Dept. of Geography, University of Waterloo, Waterloo, ON, Canada, 2013; 105p.
104. Wood, P.J.; Armitage, P.D. Sediment deposition in a small lowland stream–management implications. *Reg. Rivers-Res. Manag.* **1999**, *15*, 199–210. [CrossRef]
105. Uunila, L.S. Effects of River Ice on Bank Morphology and Riparian Vegetation along Peace River, Clayhurst to Fort Vermilion. In Proceedings of the 9th Workshop on River Ice, Fredericton, NB, USA, 24–26 September 1997; pp. 315–334.
106. U.S. Fish and Wildlife Service. *Species Status Assessment for the Furbish's Lousewort (Pedicularis furbishiae), Version 1.1*; Ecological Services Maine Field Office: East Orland, ME, USA, 2018; 59p + 3 appendices.
107. Marsh, P.; Hey, M. The flooding hydrology of Mackenzie delta lakes near Inuvik, N.W.T. Canada. *Arctic* **1989**, *42*, 41–49. [CrossRef]
108. Prowse, T.D.; Lalonde, V. Open-Water and Ice-Jam Flooding of a Northern Delta. *Hydrol. Res.* **1996**, *27*, 85–100. [CrossRef]

MDPI

St. Alban-Anlage 66

4052 Basel

Switzerland

Tel. +41 61 683 77 34

Fax +41 61 302 89 18

www.mdpi.com

Water Editorial Office

E-mail: water@mdpi.com

www.mdpi.com/journal/water